Gender, Behavior, and Health

Gender, Behavior, and Health

Schistosomiasis Transmission and Control in Rural Egypt

Samiha El Katsha and Susan Watts

The American University in Cairo Press
Cairo New York

The American University in Cairo Press
113 Sharia Kasr el Aini, Cairo, Egypt
420 Fifth Avenue, New York, NY 10018
www.aucpress.com

Dar el Kutub No. 5377/02
ISBN 977 424 728 0

Designed by AWH.Hughes
Printed in Egypt

Contents

Figures

Tables

Foreword

This book is a clear demonstration of the role of research in guiding policies, the value of multidisciplinary research, and the importance of going beyond vertical and narrowly defined interventions.

The book moves from describing the burden and realities of a public health problem to suggesting prevention and curative approaches. It occupies itself with a health issue in a way which reflects a clear appreciation of the link between health, development, environment, and gender. It builds on a holistic, interdisciplinary approach that recognizes the social and contextual forces shaping the management and production of health.

Schistosomiasis in Egypt is a serious health hazard that has been defying the many attempts to control it. The authors of this book are one step toward this goal. They argue for the importance of recognizing the behavior and gendered social interactions occurring during disease transmission as well as during diagnosis and treatment. They propose a model of interventions that is conscious of the realities of rural communities, that allows for the existing limitations of health services and that builds on community participation and strengths.

This down to earth and practical approach is not easy to adopt. It requires a high level of sophistication and cumulative knowledge and experiences. The authors of this book have devoted a considerable number of years researching and studying health, water and sanitation issues in Egypt. Their expertise on these issues is internationally well known and their insights are valued by academics and practitioners alike.

The Social Research Center of the American University in Cairo is pleased to have supported the work of Susan Watts and Samiha El Katsha. For us, this work embodies our mission of "contributing to knowledge in the social sciences and informing policy formulation and implementation." We are proud of their contribution and the quality of their work.

I have truly enjoyed reading this book and I am sure you, the reader, will share my enthusiasm for it.

Hoda Rashad
Director, Social Research Center
The American University in Cairo

Preface

This book is based on the results of an in-depth study of schistosomiasis in the Nile Delta carried out between 1991 and 1997. The project began with a focus on the impact of safe water and sanitation on infection in two rural communities, which we have called al-Garda and al-Salamuniya, with a third village added later. Nowadays each community has upwards of eight thousand inhabitants, although in common parlance they are still termed 'villages.' As the study developed, the researchers sought to gain an overall picture of schistosomiasis in these localities, using the concepts and methods of the social sciences and incorporating, where needed, those of public health, epidemiology, and environmental studies. Thus this is a holistic, interdisciplinary approach that combines all aspects of schistosomiasis in the community setting.

The study focuses on *Schistosoma mansoni*, the intestinal form of schistosomiasis, which now predominates in the Nile Delta. It was designed to involve the participation of all those in the villages who were concerned with schistosomiasis: health providers working in local health facilities, school teachers, village leaders, and the community residents themselves. We wanted to get a view of schistosomiasis from the grass roots level, to find out what was actually happening on the ground, rather than accepting what people in government and administration thought was happening.

This study is unique in that it looks at *all* facets of the disease in the study settlements, both preventive and curative. It is a study of behavior and gendered social interactions occurring during disease transmission at the canals, during other water- and sanitation-related activities, and during diagnosis and treatment at the local health facilities. All explorations were designed to obtain information and insights that could be used in the design and testing of strategies to improve the control of the disease at the community level. We attempted to identify control activities that could be implemented within the structure of the existing health services as well as other relevant local services, such as education, water, sanitation, and irrigation, making the best use of the available resources and staff at the local and district level.

For many years the authors of this book, a social anthropologist and a health geographer, have shared an interest in health, water, and sanitation issues. As social scientists we have carried out studies of health in relation to the daily activities of women, men, and children, which have centered around the use of water in communities where safe water and functioning sanitary facilities are not always found. We have attempted to ascertain local perceptions of health, and the strategies involved in the search for treatment. Our ideas about schistosomiasis in the rural and semi-rural setting evolved and matured during the period of study. As we worked with our professional colleagues and local community members, we became increasingly aware of the extent to which our participatory approach helped us understand gender issues.

This book is designed for readers with a general interest in health, environment, development, and gender issues. In the first part we present the general background of schistosomiasis and the situation in Egypt in the early to mid-1990s. The second part of the book focuses on our in-depth community-based study and proposed interventions to help to control schistosomiasis.

We hope that the presentation of our research findings is convincing for specialists and for non-specialists, and that these also reflect the 'reality' as understood by the community members, health providers, teachers and other local participants in this research project. We would like to think that our messages about gender sensitive interventions and activities and community participation will encourage policy makers and planners, as well as those living and working in village communities endemic for schistosomiasis, to think creatively about more effective strategies for the prevention and treatment of the disease.

Papers published by the team members in professional, peer-reviewed journals are listed in the bibliography, along with other sources. Three of our papers have been translated into Arabic and are available, on request, from the authors at the Social Research Center (SRC) of the American University in Cairo. These papers are: "The public health implications of the increasing predominance of *Schistosma mansoni* in Egypt"; "Changing environmental conditions in the Nile delta: health and policy implications with special reference to schistosomiasis," and "Community participation for schistosomiasis control: a participatory research project in Egypt".

During the research period, team members presented their findings at conferences and workshops, especially at the International Conferences on Schistosomiasis sponsored by the Schistosomiasis Research Project in Cairo in 1991, 1993, 1995, and 1998. They helped to organize workshops on women and gender issues at these conferences in 1993 and 1998, and a round table session on health education in 1995.

At the 1995 meeting, Samiha El Katsha presented a plenary session entitled: "Old ways may be engrained: an anthropological perspective on schistosomiasis." She presented a paper on the social aspects of schistosomiasis at the conference of the American Society of Tropical Medicine and Hygiene meeting in Atlanta in November 1993. She has also been actively involved in activities related to GARNET (Global Applied Research Network) and the Water Supply and Sanitation Collaborative Council (WSSCC).

In January 1999, Susan Watts prepared a review on "Gender and Schistosomiasis in Egypt" for the Task Force on Gender Sensitive Interventions, of the WHO Special Programme for Research and Training in Tropical Diseases. She spent three months, February–April 2000, as a visiting researcher at the Key Center for the Study of Women's Health in Society, at the University of Melbourne, Australia. These various activities helped the authors to crystallize their views about gender and schistosomiasis, and environment and health, and strengthened their determination to write about these issues for a wider audience.

Acknowledgments

The research project on which this book is based was funded by the Schistosomiasis Research Project, under research grant agreement # 04-05-38, which lasted from September 1991 to February 1997. It was conducted under the auspices of the Social Research Center of the American University in Cairo. We are grateful to successive directors of the Center, Dr. Saad Nagi and Dr. Hoda Rashad, for their support and encouragement. At AUC, we also thank successive Directors of Sponsored Programs, Dr. J. Collum and Dr. Steven Goode, and the Associate Director, Mouna Shaker. Thanks are also due to Professor Gilbert White for funds to support the summer clubs in Munufiya.

The team is grateful for the support of Dr. Taha El Khoby, Director of the Schistosomiasis Research Project and, until his retirement in 2000, First Undersecretary in the Ministry of Health and Population, and to Dr. Alan Fenwick and Dr. Nabil Galal, respectively Project Manager and Technical Director at the Schistosomiasis Research Project. They encouraged us throughout the research process and during the preparation of this book.

The research team consisted of Samiha El Katsha, a cultural anthropologist, Susan Watts, a geographer, Awatif Younis, a sociologist, Esmat Kheir, a sociologist, and Hanan Sabei, an anthropologist—all from the Social Research Center of the American University in Cairo—and Amal Khairy, an epidemiologist and public health specialist, Olfat Sebei, an environmental chemist, and Osama Awad, malacologist, from the High Institute of Public Health in Alexandria. Eqbal Sami, Hind Al-Helaly, and Gunilla Soliman at SRC provided technical and logistical support. We thank Dr. Barnett Cline and other schistosomiasis specialists in Egypt and the United States for support and encouragement during the research project, and Dr. Lenore Manderson for inviting Susan Watts to spend three months at the Key Center.

We are grateful for support from the Health Directorate in Munufiya, especially our local consultant, Dr. Salina Anestasi, former Director of the Endemic Diseases Unit, and the snail control specialist, Engineer Mohamed Atwah. Village support was provided by school headmasters Samir Hendawi and Kamel Badawi and local teachers who volunteered to work on the project; also health unit staff, especially the director, lab technicians, nurses, and the school health nurse. Finally, our thanks are extended to villagers and all local residents and village leaders who participated in this study, without whom it could not have been carried out.

The two authors are grateful to Dr. Hoda Rashad, director of SRC, for providing for a grant from ENRECA, a Danish aid program, toward support during the writing of the book and for help in the preparation of diagrams and maps, so ably prepared by Moody Youssef. We thank Neil Hewison, Mary Knight, and Kelly Zaug at the American University in Cairo Press for support, advice, and encouragement. We are also grateful to Sheldon Watts for editorial vigilance.

Acronyms

CDA—Community Development Association
EDHS—Egypt Demographic and Health Survey
EHDR—Egypt Human Development Report
MOH—Ministry of Health
MOHP—Ministry of Health and Population (in Egypt after 1997)
NGO—Non-Governmental Organization
NOPWASD—National Organization of Potable Water and Sanitary Drainage
NSCP—National Schistosomiasis Control Program
RHU—Rural Health Unit
SRP—Schistosomiasis Research Project
TDR—UNDP/World Bank/WHO Special Programme for Research and Training in Tropical Diseases
UNDP—United Nations Development Program
UNICEF—United Nations International Children's Emergency Fund
WHO —World Health Organization

Glossary

'izba (pl. *'izab*)—hamlet
hosh—courtyard
markaz—district
misqa—field canal
mikrubat—microbes
muwazzaf/a (pl. *muwazzafin*)—government employee
qaria (pl. *qura*)—village
qaria umm—'mother village'
saqia—waterwheel
wihda sihhiya—rural health unit
zimam—fields around the village

During our fieldwork (1992–96), 1 Egyptian pound (100 piasters) was worth about US$0.30 (US$1 = 3.3 Egyptian pounds).
One feddan = 0.42 hectares = 1.038 acres

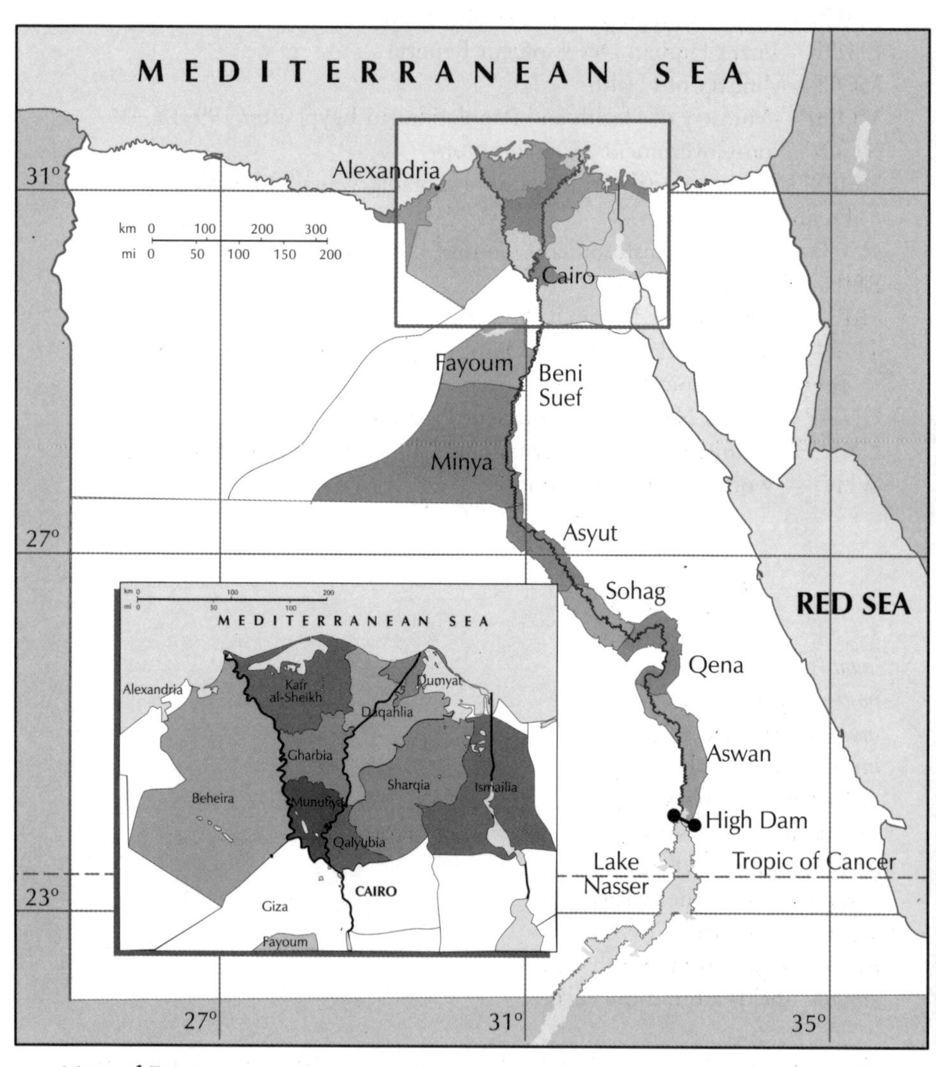
MEDITERRANEAN SEA
Alexandria
31°
km 0 100 200 300
mi 0 50 100 150 200
Cairo
Fayoum
Beni
Suef
Minya
27°
Asyut
Sohag
RED SEA
Qena
Aswan
High Dam
Lake
Nasser
Tropic of Cancer
23°
27°
31°
35°
MEDITERRANEAN SEA
Alexandria
Kafr
al-Sheikh
Dumyat
Daqahlia
Gharbia
Beheira
Sharqia
Ismailia
Munufiya
Qalyubia
CAIRO
Giza
Fayoum

Map of Egypt

1

Overview

Why study schistosomiasis and why in Egypt?

Schistosomiasis is a parasitic disease transmitted through human activities taking place in or near canals, slow-moving streams, and lakes in tropical and subtropical countries. It is especially prevalent in Africa—including Egypt—and in Brazil and China. As of the mid-1990s, approximately two hundred million people worldwide were infected. Three times that number, six hundred million, were estimated to be at risk, mainly in poor rural and semi-rural communities with limited or no access to safe water and sanitation. When schistosomiasis victims first become infected they suffer intestinal disorders, nutritional deficiencies, and general debilitation. Serious long-term effects occur when internal lesions develop into cancers, or when malfunctioning organs cause blockages elsewhere in the victim's system, leading for example to enlarged spleens and livers. In global terms, around 10% of those infected with schistosomiasis (around 20,000,000 people at present) go on to develop serious health problems (Chitsolu et al. 2000).

People who enter a canal for occupational or recreational purposes are at risk of infection if the water contains schistosome larvae (we will explain this process at greater length in chapter 3). After these late stage larvae penetrate the human host, they migrate to the liver where they mature within six weeks, to worms about one centimeter in length. They then pair (male and female) and migrate to the blood vessels around the bladder (in the case of *Schistosoma haematobium*) or intestine (in the case of *Schistosoma mansoni*). From their resting sites, the female worms produce eggs which escape from the blood vessels into either the bladder or intestine. They leave the infected person in the urine or stool. These eggs reach the canals or other surface water as a result of the improper disposal of latrine effluent or of indiscriminate defecation or urination. Therefore strategies to prevent the infection require investigation on two levels. At the local level, prevention strategies involve an understanding of the behavior of people living in these at-risk communities and the water supplies they come into contact with on a daily basis for domestic purposes, for irrigation, and for recreation. Prevention strategies also require an examination of the available domestic water supplies, sanitation, and drainage, and the upgrading of these facilities so as to interrupt the disease cycle. The upgrading of infrastructure takes us beyond the local community level, to a consideration of the tasks and responsibilities of government at the local, district, and ultimately, at the national level. These tasks and responsibilities will be one of the major aspects of schistosomiasis that we will explore in this book.

Schistosomiasis has been recognized as a serious health problem in Egypt for the last one hundred and fifty years, although infection levels fell sharply during the 1990s. People living in rural areas where the disease is most often found comprise about 50% of the country's total population of around 65 million. As of the late 1990s, it was estimated that 10 to 20% of these people were infected (Doenhoff et al. 2000). The far larger numbers of people who continue to be at risk of infection are those who come into contact with the waters of the irrigation canals that flow through the fields and densely settled rural communities in the narrow river valley south of the capital, Cairo, and in the broadening Delta to the north.

Egypt, although a low income country, has a nation-wide public health care system that reaches the vast majority of the rural population in these irrigated areas. Both the Delta (sometimes known as Lower Egypt) and the narrow Nile valley south of Cairo (Upper Egypt) are included in a National Schistosomiasis Control Program, initiated by the Ministry of Health in 1977 and delivered through primary health care centers. This Control Program has, in practice, two major approaches. One is the treatment of affected individuals, a strategy designed to break the chain of disease causation by destroying the parasite and its eggs in the human host. The second approach is to kill the snail host external to the human, by applying chemicals to affected canal water. In 1988, Egypt was the first country to make praziquantel, a new oral drug that could successfully treat most (but not all) cases with a single dose of tablets, available free to all through government supported health services. One of the subjects we will explore is what happens in rural health facilities when patients come in for diagnosis and treatment.

The social science perspective

Diagnosing and treating infected individuals and destroying the snail vectors are not the only ways to control schistosomiasis however. From our point of view, as social scientists, these approaches underestimate the complexity of human behavior. Social scientists look at infection as the result of daily activities carried out by various groups of people in specific social settings. As such, we view schistosomiasis as a disease of human behavior, and our understanding of this behavior begins with the recognition that it is not random, but socially patterned (Huang and Manderson 1992).

Social scientists also look beyond immediate happenings at the local level to examine the broader political and economic structures that form part of the greater whole within which the immediate events occur. We need to ask how national and regional policies are translated to activities in the localities, and to identify possible structural barriers to activities designed to control disease. Unfortunately, many health activities are planned at the national level with little scope for taking account of local conditions. Understanding what actually happens at the local level requires anthropological inquiries into local knowledge and local practices (Manderson 1998).

At the local level, anthropologists probe into the question of why specific social groups continue to use the waters of the canals, even though they know that through this water contact they run the risk of contracting schistosomiasis. They will expect to find that members of different groups, defined by age, gender, and the task involved, will provide a variety of different answers. Farmers, for example, may demonstrate that it is impossible for them to carry on with irrigated farming, as now constituted, without coming into frequent contact with canal water. Young single and married women, on the other hand, will have quite a different explanation of why they continue to use the canal for doing laundry or other domestic tasks.

It is also important to understand behavior at another point in the transmission cycle, when humans contaminate water sources. Differences in excretory behavior are likely to be associated with the two types of schistosomiasis found in Egypt, *Schistosoma mansoni* in which eggs are excreted in the feces, and *Schistosoma haematobium*, in which eggs are excreted in the urine (see chapter 3). Here too, we need to consider both local water and sanitation facilities and their use, and the prospects for the provision of safe latrines, safe disposal of latrine effluent and a safe drainage/sewerage system that will provide for the needs of the entire population.

For the social scientist trained as a geographer, the interrelationship between social networks and patterns of behavior, and the physical setting (in this case, that associated with various aspects of water supply, water use, sanitation, and irrigation) is the core of the discipline. In this sense, geography is unique in that it integrates the various elements that occur together in a single location. The gendered behavior associated with exposure to schistosomiasis also occurs in specific *places*, at or very near canal water. Distance is also an aspect of place. For example, for a woman living in a house without an individual water connection, the distance between her house and an alternative water source—a shared tap, handpump or canal—can influence her choice of water for domestic purposes. Similarly, the distance between a residence and a health facility may affect a person's decision whether or not to use that facility.

When medical anthropologists study ill-health, they use the term "illness" in a specific sense, to identify the patient's interpretation of their condition. In contrast, in the biomedical view, "disease" focuses on the disease agent (the pathogen) and its impact on the human body. In clinical practice, when meeting a patient a physician is interested in how the individual patient's suffering can be relieved. At the professional level, she or he is less concerned with how the disease was acquired.

Epidemiologists, who usually identify their discipline as a subdiscipline of biomedicine, focus on identifying the proportion of those with a disease, and the mortality and sickness it causes. As researchers, they are concerned with a whole population, rather than with an individual. They differ from social scientists chiefly in their unwillingness to pursue the question of *why* people become infected (Inhorn 1995).

Throughout our research project, we have drawn on the findings of biomedical researchers, especially epidemiologists. Members of our interdisciplinary research team have supervised an epidemiological survey in the study communities, and an environmental study of water quality and snail populations in canals. This led us to consider the overall characteristics of the canal systems in the two study communities that might hinder, or encourage, the survival of the vector snails. Thus, in our research practice, we viewed these diverse disciplinary approaches as complementary, rather than as being in opposition to each other.

Studying gender

One of the central purposes of this book is to explore the issue of gender and gender roles in community settings. In current social science literature, gender "denotes socially constructed behaviors, expectations and roles that derive from, but may not depend on sex" (Vlassoff and Manderson, 1998, 1011). Gender, as it relates to schistosomiasis, focuses on the social, cultural, economic, and behavioral attributes of males and females that relate to transmission, diagnosis, treatment, and prevention. As such, gender is reflected in day-to-day activities relevant to disease transmission—such as domestic water use and disposal, and canal contact—as well as to the interpretation and treatment of illness.

Biological differences are also relevant as an infection often affects women and men differently. For instance, it may produce distinctive symptoms or tissue damage, and these distinctive signs and symptoms have a socio-cultural significance. They affect the way in which women, in their own way, and men, in theirs, interpret the severity of their own illness and that of other family members. The interpretation of these symptoms also influences the decisions family members make about when and where they should go for diagnosis and treatment.

Women and men might have very distinct understandings about their state of health or ill-health. For example, from a biomedical point of view, the amount of disease clinically identified in women and men may not be very different. However, women may be more sensitive to bodily clues related to disease and therefore suffer more "illness" than men. But because of their role assignments and cultural expectations, fewer women than men who feel "ill" may interpret this as requiring that they seek outside health/medical care. For example, many women in low income countries who endure a continual and heavy daily round of domestic and farm work do not want to be seen as "weak" by their husbands and senior female relatives. They fear that an admission that they need medical care will threaten their central role as wives, providers, and carers for their family. Thus they are often described as stoically enduring illness that would send sophisticated city-bred women straight to bed. For these reasons, even though they experience more "illness" than men, women are less likely to seek care from a formal health provider than are their menfolk (Rathgeber and Vlassoff 1993).

This book presents a case for introducing greater awareness of gender issues into all future health interventions. Adopting a holistic approach, it will be seen that men and women have distinctive roles and expectations that are complementary, even if at the same time they are sometimes in conflict. Except in specific cases, focusing exclusively on females is seen as presenting an unbalanced view of the depth and richness of social processes.

Studying a community

Social scientists have expended a great deal of time on the question: "What is a community?" We no longer think of a community in romantic terms as a group of people sharing common interests and viewpoints that can easily be identified (Whyte 1983). For us, at a very basic level, the "community" is best understood as a cluster of people of all ages and of both sexes who share a common, bounded geographical space.

A community itself may have doubled in size in the last twenty years (in common with Egyptian-wide demographic patterns), yet it remains a cluster of people who use local schools and whatever health facilities are available. The members of a community also experience common environmental problems such as poor sewerage/drainage or facilities for disposing of sewage effluent. As such, a community is also a location for health-related activities and potential environmental improvements that will benefit all residents.

Community residents may also hold widely conflicting views. Thus in the two Nile Delta communities which we studied, most people shared a common perception of schistosomiasis as a disease which caused ill-health. However, their knowledge base about the disease in its various stages in a human host, in a snail host, and so on, differed widely. Some had personal experience with the disease, others did not. These differences affected their ideas about how to avoid becoming infected in the future and how to prevent the infection from being transmitted to the wider "community."

This mix of ideas, born of lived experience in a rural setting with its distinctive water use requirements and practices, obviously differs from that of desk-bound middle-class Ministry officials in the regional capital or in distant Cairo. For us, the significance of community-based research in health is that it attempts to look at the situation reflexively from the vantage point of the various groups within the community itself. As social scientists, we have a view of schistosomiasis that focuses on human behavior and social interactions within a community setting. We are convinced that this approach can provide useful insights for planners at the regional, national, and international level. We turn now to our particular conceptual approach: action-oriented research.

Action-oriented research

The objective of "action research" is to involve local people in activities to improve their own well-being. In our study we worked closely with ordinary men and women in the community, with local health providers, and with school teachers, in the collection of a great range of information about schistosomiasis in the local setting. Thus, at the same time that we are interested in identifying relevant information, we are also concerned with "capacity building"—increasing skills and upgrading formal and informal institutions (Cline and Hewlett 1996; El Katsha et al. 1993–94).

Ideally, the process of action research in health begins with local people's own perception that they have a health problem. Given to understand that what they said about the community's problem with schistosomiasis would be listened to, and perhaps incorporated into disease control initiatives, many categories of people contributed to the on-going dialogue: community leaders, women, health providers of various types, and school teachers. All of these partners (or stakeholders, as they are sometimes called) were encouraged to draw on their special knowledge and insights in contributing to the task of improving community health (see Nichter 1984).

Action-oriented research cannot claim to be "disinterested." And because of its nature as a flexible, open-ended exploratory process, responsive to the suggestions of all partners, neither can it be expected to follow a set of predetermined procedures. The purpose of action-oriented research is not to obtain a specific body of information. Instead, it is to come up with new insights that can be built into the next stage of the inquiry, in open-ended fashion.

Researchers using action-oriented techniques do not see themselves as casual visitors dropping in on a community to administer sets of questionnaires that were designed to cover any and all rural conditions. Instead, participatory action research requires sustained contact over a relatively long period of time. The use of local people and health staff to collect information helps to ensure a frank and open exchange of views about such things as what they know about schistosomiasis and why they use the canals. It also requires that there be a considerable amount of feed-back between the initiating research team and all the partners in the project, as well as with the wider community itself. In this context, the researchers reject any claims that they are the only "experts," for they know full well that the knowledge of local people is essential to the success of the enterprise (see Chambers 1997).

The arrangement of the book

The book is arranged in two parts. The first chapters, 1 to 5, present general issues relating to gender, health, and the environment, and provides a background to the global and Egyptian context of schistosomiasis. The second part of the book presents the specific findings of our research, beginning with research methodology and procedures (chapter 6), and the village social and environmental setting (chapters 7 and 8). Tracing the stages of the disease in humans, chapter 9 looks at human behavior at canals that exposes people to infection, and chapter 10 looks at the villagers' view of schistosomiasis, how they become aware that they might be infected, and their view of what they should do next. Chapter 11 explores what happens when they go to a local Rural Health Unit for diagnosis and treatment, and what could be done to improve diagnosis and treatment for adults, as well as for school children.

Chapter 12 looks at community-based approaches to control. The final chapter, 13, begins by summarizing the approach and findings of our study. It looks ahead to the situation in the first years of the new millenium, exploring the relevance of our findings about environmental changes and about health provisions.

2

Gender Issues in Rural Egypt

The rural setting

In this chapter we take up the theme of gender in the rural environment as a starting point for our study of schistosomiasis in the community setting, focusing on the years of our study, 1991 to 1996. We start with the recognition that males and females do different things at different times in their life. Yet, as a general rule, everyone who lives in rural Egypt—whether it be in the Delta or the narrow river valley south of Cairo—who acquires and transmits the disease does so within their own community, and seeks treatment at local health facilities. In these activities, they are guided by their knowledge of the disease, and by the roles and responsibilities that their family members and neighbors expect of them.

The links between gender and health are played out in the Egyptian setting against a background of rapid political, social, and economic change. The Revolution of 1952 promised a better life for all citizens. It led to the establishment of legal rights to free health care and education for all, and gave women equal rights in employment and political participation. To achieve these objectives the government began to expand health and education infrastructures. Then, beginning in the mid-1970s and continuing today, *infitah*, the process of "opening up" and privatization, allowed for the extension of private health services and education, to supplement those provided by the state.

Meanwhile, the size of the population continued to grow, from 27.9 million in the 1966 census to 64.7 million in 1996. By 1997, life expectancy at birth had increased to 66.3 years, compared 50.9 in 1970 (UNDP 1999:

170). The doubling of population every twenty years has been due to decreasing rates of infant mortality combined with a large number of births; although the birth rate per mother has declined there are still a large number of young adults—future parents—in the total population. This population growth put strains on the infrastructure, the environment, and the capacity of the economy to provide jobs (especially for the increasing number of young people).

By the early and mid-1990s, the majority of Egyptians were better off, in material terms, than they had been forty years earlier. Basic foodstuffs and transportation were now subsidized by the government. However, what it meant to be poor had changed over time. In 1995, a study found a growing disparity in income distribution between the richest and the poorest twenty percent of the population, and the persistence of extreme poverty in some rural areas (especially in the south) and in some urban neighborhoods in Cairo and Alexandria.

At the same time, there was considerable evidence for increased upward mobility in education and occupation for young women, compared to their mothers, and for young men, compared to their fathers (Nagi 2001: 238, 241). However, actual improvement did not always keep pace with the even more rapid rise in expectations.

As a result of all of these changes, farming lost its role as the economic mainstay of many rural communities, especially in the Nile delta. However, many men and women continued to farm on a part-time basis, to supplement their income from government employment. Thus these farmers, whether full-time or part-time, men, women, or children, continued to be at risk of contracting schistosomiasis because of their exposure to canal water possibly infected with vector snails and schistosomes.

As of 1991, 53% of Egypt's population was considered to be "rural." Since the introduction of local government reforms in 1960 to bring services to rural residents, this has been an administrative definition, embracing all settlements that are not regional (governorate, *muhafza*) or district *(markaz)* headquarters. This new structure identified a "mother village," *qaria umm*, for each village area consisting of a main village, half a dozen or so satellite or sub-villages *(qaria tab'a)* and a varying number of hamlets *('izab)* scattered in the fields. Situated in the mother village is the Village Council, administered by the Ministry of Local Government. This Council has two parts, firstly the Executive Village Council consisting of permanent civil servants responsible for maintaining local services such as water and sanitation, as well as overseeing agriculture, health and the supply of subsidized commodities. The elected Popular Village Council, consisting of residents of the village area, men and women, acts as a channel for requests for services from communities within the village area (El Katsha and Watts 1993: 18).

Rural residents think of themselves as living in villages, *qura* (pl. of *qaria*), and compare themselves to residents in towns and cities (Hopkins and Westergaard 1998: 4–6). To outsiders, these villages, which may have populations of over 10,000 people, look urban, with densely packed houses, schools, government offices, shops, and lots of activity in the streets. This impression of dynamism is increased by the number of new, completed, and partly-built houses that crowd the approach roads to the villages.

In general, socioeconomic conditions and the position of women have been less favorable in rural than in urban areas. This is indicated by lower levels of access to safe water and safe sanitation, and the smaller proportion of educated women and of girls in school in rural compared to urban areas, contrasts we will examine later in this chapter. However, within the rural areas, the Nile delta—the site of the settlements studied in this book—enjoyed better conditions than did most of the rural areas in the river valley south of Cairo (including the governorate of Fayoum, a depression west of the valley, fed by the waters of the Nile).

In the Delta, as elsewhere in rural Egypt, continuing change is associated with such things as fluctuations in the pattern of migration to Egyptian cities and overseas, as well as increased access to transport, and to both state and private health and education facilities. All these changes are taking place against a background in which rural Egyptians are becoming increasingly aware of the rest of Egypt and the rest of the world, thanks to television and its urban bias.

In the midst of all these changes, issues of gender are complex and can be looked at from a number of different angles. From the point of view of our study, gender relationships are best explored in terms of what women and men, girls and boys, actually do in their daily life and the reasons they give for their behavior. Actual behavior, and an individual's rationale for it, do not always coincide with what anthropologists call "gender ideologies"—the normative, collective view of what should be done. Finally, there is a gap between the legal status of women and their actual access to education, health facilities, and political power (see, for example, Moore 1988: ch. 2).

A national dialogue

The discussion of gender in Egypt can be seen in the wider context of the international dialogue about gender and the family. In 1994, the International Conference on Population and Development, held in Cairo, played an important role in broadening the scope of the debate, by beginning to consider gender rather than purely women's issues. It also stressed the need to involve non-governmental organizations (NGOs) in strategies to improve rural life, and especially the position of rural women.

The approaches introduced at the Cairo conference were consolidated at the Fourth World Conference on Women, held in Beijing in September 1995. Representatives from Egypt included nationally known figures, as well as a group of women from a rural based NGO who had not previously traveled outside their home district. These women, like their sisters elsewhere in Egypt, needed a vocabulary to articulate their awareness of gender as it affected Egyptian women.

Yet, the Arabic language, in its classical form (the language of the Qur'an) and in its colloquial form (the everyday language of the people), has no single word for "gender" as distinct from "sex" (male and female). Some international agencies working in Egypt use the term *gendera*. This is an obvious European import and had a discordant ring in spoken Arabic. More importantly, the word *gendera* seems to imply that there are no differences in the nature of gender issues in Arabic-speaking lands and gender issues perceived to be important in the West, where the most strident forms of feminism arose, and where marriage to someone of the opposite sex is no longer seen as essential to the maintenance of social identity.

Middle Eastern women were well aware of this problem at the Cairo and Beijing conferences. Since then, a search has been going on for a more appropriate term to express their aspirations. A recent magazine article suggested the words *al-nua al-igtimai*, loosely translated as "the social aspects of sexual identity," as it focused on the importance of the roles and responsibilities of women, men, and children. This definition was seen as being more appropriate than other terms that could be translated into English as "equity between the sexes," *al-musawa al-igtimai*, or "equality between the sexes," *al-'adala al-igtimai*. Although the idea of "equity" is more acceptable to Egyptian ways of thinking than "equality," both could be seen as contentious in the Middle Eastern context.

Egyptian government social policy in the late twentieth century increased its concern for the well-being of women and children. In 1980, it declared a Decade for the Protection of the Egyptian Child. This initiative was followed a short time later by the establishment of the National Council for Childhood and Motherhood and the National Council for Women (NCW), thus preparing the way for a dialogue about gender.

The goal of the NCW was to empower women "to play an effective role in the social renaissance Egypt is currently undergoing. It also aims at enabling women to better preserve their national heritage and the Egyptian identity and raise generations of Egyptians harboring love for their homeland and holding on to their traditions and religious institutions." These goals stressed the role of women as conservators of heritage, as guardians of social values and mores, and as fulfilling roles which complemented those of men.

But the NCW also saw the need for practical measures to support family health, such as improving access to safe water and sanitation, and achieving total coverage for vaccination against childhood diseases. These uncontroversial aspects of the NCW program undoubtedly could be accepted by the majority of Egyptian women and men. However, other parts of the NCW mandate were more radical. For example, these included organizing "awareness-raising training sessions on women's rights, duties and role in society" and drawing up a national plan for the advancement of women (NCW 2000/1).

In the late 1990s, the debate about family issues and gender roles centered on proposed changes in the personal status law, *al-ahwal al-shakhsiya*, that stipulated the conditions for divorce and child custody. Some men and women supported the right of a wife to initiate divorce, and to work outside the home and travel abroad without her husband's permission. Others opposed these changes on the grounds that they would threaten the stability of the family.

The Egyptian family: continuity and change

Gender—understood as women's and men's roles in the family and in society—is a common theme in national debate, widely discussed in public, and in the mass media in newspapers, magazines, and television. This constant debate reflects an awareness of the uncertainty that surrounds the social relationships of women and men, and the rules that should guide these interactions.

Especially in rural communities, a structure of power and deference based on well-defined gender rules is still the basis of interactions in the family and in the community. A statement about life in a conservative Egyptian hamlet in the mide 1980s illustrates why gender issues have been so fiercely contested over the last twenty years:

> Women and men are addressed and address each other differently. Roles that women and men enact are well defined and separate, and children's role cues are taught explicitly, precisely and with conviction. These rules cover all aspects of daily life (Lane and Meleis 1991: 1199).

It is these rules that were being challenged in the 1990s.

In general, one can state that men have higher status than women, that older men have power over younger men, and that older women have power over younger women. Within the household, the shared residential setting for family life, women's and men's roles and activities depend on their position in the life cycle. The senior resident male has unquestioned power over all members of the family. He is responsible for directing any farming activity the family might be engaged in, and he controls expenditure outside the immediate domestic sphere. A mother with adult sons has power over her daughters-in-law. The senior woman is responsible for the distribution of domestic and economic tasks, where and when they should be done, and by whom.

This pattern of role assignment is most clearly in evidence in large extended families consisting of co-resident parents and one or more married sons and their wives and children. However, it can also be seen in the two-generation nuclear family that is rapidly becoming the norm among prosperous urban Egyptians (many of them sons and daughters of first generation immigrants from the countryside).

Given the recognition of the power of the senior male over other family members, census findings that between ten and fifteen percent of Egyptian households are headed by women would seem to require an explanation. For some households, the husband is only temporarily absent, working elsewhere and often sending money home. In other households, the woman has been left alone due to widowhood or divorce. Yet here, being "alone" usually means that a male kinsman has power over her activities, and is the guardian of her children. Especially in rural areas, most widowed women are under the authority of a senior male from her husband's family or from her own family of birth.

Many women who in census terms are "heads of households" (either temporarily or in the long-term) are at a disadvantage economically. A recent study of poverty in Egypt, which identified 15.7% of sample households as being headed by women, found that these households were at a comparative disadvantage in terms of per capita income and expenditure as well as basic needs (for food and clothing). The study found that in rural areas, female heads were mostly found among non-farming households (Nagi 2001: 61–64).

Kamran Asdar Ali (1998) comments that some aspects of the dynamics of family life have often been overlooked because of anthropologists' and feminists' stress on patriarchy and the low status of women. In his study of a village in Sharqiya governorate in the eastern Delta, he mentioned the benefits of the privacy and companionship that both women and men enjoyed when they lived in small nuclear family units. Their separation from parents and siblings encouraged the growth of affection between them. The corollary of this was that women whose husbands were working away felt a real sense of loss, as well as being aware of their additional load of work and responsibility. In this Sharqiya village, in recent years both women and men appreciated their increasing freedom to choose their spouses and to establish separate households, a goal which many had achieved while still respecting the opinion of their parents.

In Egypt, one remarkable change has occurred in the last couple of decades that is bound to have an impact on the structure of the family: the age of first marriage of women as well as of men has substantially increased. The 1995 Egypt Demographic and Health Survey (hereafter EDHS), based on a national sample of urban and rural households, found that only one quarter of Egyptian women aged 20–24 were married by the age of 18, whereas half of women in their parents' generation (aged 45–49 in 1995) had been married at that age. The median age at first marriage for urban women aged 24–29 was 21, just over three years higher than for rural women (EDHS 1995: 115).

Women in rural areas marry earlier than those in urban areas, and in Upper Egypt earlier than in the Delta. In 1992, in a village in the Delta governorate of Sharqiya, 30% of women said that they had married before the age of 16, and 74% before the age of 20. The corresponding figures for a village in Aswan, in southern Egypt, were 60% and 89%. Women who married as teenagers said that they had received little or no schooling, so "What was I to wait for?" (El Hamamsy 1994: 10, 4, 11). Most of the women who had married at twenty or above said that their later marriage age had enabled them to finish their schooling. However, in poor and remote areas, fathers considered that girls did not need an education, as their role was to help at home, and in the future, to be mothers. Such ideas favored early marriage (El Hamamsy 1994: 22).

Planners concerned about rapid population growth and the health problems caused by early child bearing have viewed an increase in the age of first marriage in a positive light. However, the situation in Egypt is complex. Deeply held ideas hold that marriage should be a universally desired state and that newly married couples will soon want to have a child, even though these expectations clash with the lived experience of many educated young people.

The 1952 revolution provided free secondary and university education for all young people who passed the necessary examinations. It also guaranteed university graduates employment in the public sector. However, half a century after the revolution the availability of job opportunities has failed to keep up with the expansion in the number of graduates. This means they now have to wait five years or more for a government post. Moreover, young people now find that inflation and the shortage of available housing has pushed up the cost of an apartment far beyond their limited financial resources.

Thus, for many educated young people, the delay in marriage is involuntary. Even in rural areas, it is not unusual for a girl with a university degree to be unmarried in her mid-twenties, three or four years after graduation. This suggests that two patterns of age at first marriage co-exist, with educated young people delaying marriage, while those with little or no schooling continue to marry at a young age.

Because of delayed marriage, the Egyptian female population has experienced a marked decline in fertility. Also contributing to this decline is the increased use of contraception to space children and to limit family size. Thus, although women continue to regard motherhood as their primary role, on average they now spend a shorter period of their lives being pregnant and looking after young children. The number of babies born alive for women aged 15 to 49 (the total fertility rate) in the 1995 EDHS was 3.65, compared to 5.28 in 1979/80. But in 1995, as in the past, fertility rates were significantly higher in rural areas than in urban areas, and among poorer, less educated women (EDHS 1995: 41–42).

Gender, health services, and health status

By the late 1970s, as reflected in the deliberations of the 1979 World Health Assembly at Alma Ata, health policy makers and planners were expressing an increasing concern for disease prevention and primary health care at the local level. This assembly formally recognized the need to focus on good health (rather than looking specifically at disease), and to base health programs on the real needs of people in both rich and poor countries. An essential aspect of this policy of Health for All was its bottom-up, rather than top-down approach. Its stress on prevention also contrasted with the then-prevailing (and still largely extant) biomedical orientation toward curative care in large hospitals. Although the term "gender" was not initially used in the declaration following the Alma Ata meeting, the concept of gender, as socially constructed and as reflecting power relationships, was central to the spirit of Health for All. When interpreted in this way (and in spite of many criticisms of the approach), we maintain that the fundamental concepts developed at Alma Ata should remain a guide to local practice.

Rural Egyptians have a range of public and private medical services available to them, although some people, especially women, do not always make use of them. In the mid-1990s, public health care facilities were physically accessible to almost 90% of the total population. In the densely populated Nile valley and Delta, the vast majority of residents were within five kilometers of one of the 2,400 rural health facilities (Institute of National Planning 1998, *Human Development Report 1997/8*, hereafter EHDR 1998: 51, 52). These units provided the general population (including school children) with free diagnosis and treatment for schistosomiasis. They also provided a range of services for children under five, such as vaccination and treatment for diarrheal diseases and upper respiratory tract infections, which were the main causes of infant deaths. As is usual in countries with a rapidly growing population, however, the resources of the government health system were under pressure.

In 1996–97, only 1.5% of public expenditure (as a percentage of Gross Domestic Product) was spent on health, compared to 5.5% on education (EHDR 1998: 123). In response to the chronic shortage of funds, and pressure from the International Monetary Fund to institute a system of cost recovery and avoidance of "waste," some changes have been introduced into the formerly free public health system. Nevertheless, as of the year 2000, services for children of all ages, and for the diagnosis and treatment of schistosomiasis, remained free.

Since the policy of "opening up" *(infitah)* was introduced in the 1970s, private health care has become more widely available. Many physicians in the public sector became private practitioners after hours. In rural areas most of them specialize in services for women, especially pediatric and gynecological care and infertility. Moreover, pharmacists do far more than selling drugs. As in the past, they continue to provide advice and administer simple procedures such as injections and wound care.

Other health practitioners, especially in rural areas, provide services outside the formal health structure. For example, in rural areas many babies are still delivered by female traditional birth attendants, *dayas*. Informal practitioners are able to provide affordable services in a familiar setting. Some have been medically trained or have gained experience through working at the local health units. Thus villagers have a range of choices when they have to make decisions about sources of care for sick family members. Women have a central role in deciding when and where family members seek care outside the home. In a later chapter, we will explore ways in which gender roles and expectations in our study communities affect these decisions.

There is no question that Egyptian infants and children are more likely to survive to adulthood now than they were forty years ago. This is due to improvements in infant health care, nutrition, and hygiene, and the greater availability of safe water. The improvement in child health is most dramatically reflected in changing child death rates, with a fall in the infant mortality rate (the number of deaths per 1,000 children under one year of age) from 81 in 1961, to 40 in 1989, and 29 in 1996 (EHDR 1998: 53, 135). This means that a child under one year old was almost three times more likely to die in 1961 than thirty five years later, in 1996.

Egyptian statistics for the years before 1995 recorded less favorable health indicators for girls under five than for boys. This suggested that parents routinely discriminated in favor of sons over daughters and that a health gender gap existed for the under five age group. However, the 1995 EDHS recorded no significant difference in childhood immunization coverage between boys and girls, and only slightly less favorable measures of nutritional status for girls than for boys. One year later, a report, based on a national sample survey, found no gender differences in the use of health services for acute respiratory infections or in nutritional status in the under-fives. This left open the possibility that gender differentials are now beginning to show up at the age of six, when children should go to school (El Tawila 1997: Main Findings 14–15).

Gender, illness, and health seeking behavior

A different perspective on health can be provided by medical anthropologists who take a local view and explore the way in which rural people interpret "illness" and what they do about it. For example, some rural people regard certain conditions as "normal" that a clinical examination would reveal as pathological; in other words, in biomedical terms, they were indeed sick.

In Egypt it is accepted that women, as wives and mothers, are responsible for the daily health care of all family members. They are the first to recognize that a family member is unwell and they are a source of knowledge about household remedies, especially herb teas and special diets. As mothers they are especially concerned for the health and well-being of their young children.

From the mid-1980s, as part of a global Child Survival initiative (prompted at least in part by concerns expressed at Alma Ata in 1979), the Egyptian Child Survival Project promoted a series of interventions for under-fives that involved changing mothers' behaviors. For example, in the Egyptian National Control of Diarrheal Diseases Project (1982–91), mothers were taught about the signs of dehydration in their children due to diarrheal disease and how to administer an oral rehydration solution (Langsten and Hill 1994).

Going beyond that, however, planners tried to promote a sustainable health behavior that could prevent diarrhea, rather than solely resolve its symptoms (such as dehydration). They needed the expertise of anthropologists to provide insights into infant feeding patterns, mothers' attitudes and behaviors toward diarrheal disease, and mothers' rationale for their response to the disease (Inorn and Brown 1997: 48). To this end, the Ministry of Health, the Child Survival Project, and UNICEF sponsored a series of ethnographic studies in rural Upper Egyptian governorates (Sholkalmi 1990; Wahba 1990; El Hadidi 1990). These showed that mothers' understanding of their children's illness differed markedly from that of biomedically trained doctors. In addition, the studies showed how diarrheal disease was linked to poverty and the lack of safe water supplies.

As well as being mothers of children who get sick, women have their own health needs, which are often neglected. According to a study carried out in the mid-1980s in a Delta hamlet, women said that their main health problems were the result of multiple pregnancies and constant hard physical labor. They continued with their long daily round when they were in pain and distress, and when pregnant, until just before their child was born (Lane and Meleis 1991).

Recent studies in two villages in Giza and in a hamlet in Fayoum governorate found that women did not seek care for reproductive health problems for fear of censure from their husbands, due to the stigma and secrecy surrounding these problems. Moreover, infertility was regarded as a serious problem that could jeopardize a woman's relationship with her husband. In such cases, women sought care from informal healers or from private doctors, rather than at a public facility (Khattab et al. 1999; Talaat 2001).

It is important to find out how rural Egyptians currently make decisions about who in the family should seek care, and how they choose between "traditional" carers and the public or private modern sector. An anthropological study conducted in the mid-1980s in a Delta hamlet found that perceptions of who required treatment depended on the social position of the person who was recognized as being "ill." Resources of money and time were generally used for the treatment of males rather than of women or girls, regardless of the severity of the illness. Delta hamlet dwellers valued modern biomedical care, but they did not always use it because of its high cost and the cost of transport beyond the hamlet. They also felt that the educated, public sector health providers looked down on them as illiterate residents of a small farming hamlet. They considered that private providers were more responsive to their needs, even though they charged fees (Morsy 1993: ch. 5; Morsy 1978). More recently, especially in less isolated settlements, the balance appears to have tipped away from "traditional" care toward public and private biomedical care. Because of the spread of education and greater media exposure, there may now be fewer barriers to poor women seeking biomedical care.

In another anthropological study, dating from the mid-1980s, the status of the household member who was ill was seen as particularly significant. After trying household remedies for eye disease, high-status household members sought treatment from a biomedical practitioner (usually in the private sector). On the other hand, low status individuals, if they did seek treatment, tended to go to a "traditional" healer. Here, the researcher identified a status system that determined the kind of treatment people in the hamlet received: 1) among people of the same age, males had a higher status than females; 2) status increased with age and 3) class differences based on wealth, occupation, and education determined status. This last determinant identified a status difference between "peasants" and physicians, as the physicians posted to rural public health facilities usually looked down on poor people (Lane and Millar 1987).

The search for care is a complex process. For example, in a southern Delta village in the mid-1980s, a study by a member of our research team found that villagers were pragmatic in their choice of treatment, often going from one type of health provider to another. Some even visited a modern-sector physician at the same time that they were undergoing a course of treatment from a "traditional" healer. Villagers had fairly clear perceptions about which complaints were best treated by "traditional" providers—such as bonesetters, "health barbers" and *dayas* (midwives) —and which required a visit to a modern sector clinic and physician (Assaad and El Katsha 1981).

Ten years later, in the case of schistosomiasis in the two rural communities in Munufiya governorate, local people (as we shall see later in this book) regularly made use of the local health center for free diagnosis and treatment. As far as they were concerned, local remedies and "informal" sector providers had no role in treating this particular disease. Instead they held that schistosomiasis was situated firmly in the category of diseases that were treatable by modern biomedicine.

Gender, literacy, and education

The discussion on education in the *Egypt Human Development Report, 1997/8* begins with the statement that "Education is probably the most powerful instrument in achieving economic growth, reducing poverty and improving living standards" (p. 32). The report might have added that the single best investment for a country's future is education for *girls*. Epidemiologists, public health specialists, and planners have long recognized that educated mothers do more than uneducated ones to ensure the good health and survival of children who will become the parents of the next generation (World Bank 2001: 78–83). Egyptian statistics bear this out.

A study in 1980 found that in the rural Delta, the mortality rate of infants with illiterate mothers was 128 per 1,000 live births during the first year of life, compared to 68.5 for infants with literate mothers. Increasing mothers' educational level has also been shown to reduce the differentials in male and female infant mortality. This suggests that educated mothers are less likely to discriminate between their children in the provision of care on the basis of gender (Allen 1989: 46–47).

In spite of free and mandatory education for both boys and girls between the ages of 6 and 14, current school enrollment levels for girls remain lower than for boys. A national sample survey in 1996/7 found that only 72% of girls aged 6 to 10 were enrolled in school, compared to 83% of boys. Not unexpectedly, the greatest gender gap is found in rural areas, where one third of girls aged 6 to 14 had never attended school, compared to 9% of boys (El Tawila 1997: 59). Some of the reasons for this are economic. Even though school attendance is free, parents need to buy uniforms, shoes, and books, and pay incidental fees. Thus, poor parents often give priority to the education of their sons. As many health services, including the diagnosis and treatment of schistosomiasis, are delivered through the schools, this (as we shall see) puts girls at a disadvantage.

Despite this continuing gender gap in school attendance, over the long term the educational status of Egyptian women has improved. The EDHS for 1995 (p. 20) found that the median number of years of schooling among girls between 15 and 19 was 9.3, while among women between 35 and 39, roughly the age of their parents, it was only 2.6 years. Although there was a high overall dropout rate for pupils in primary schools, the rate for girls was lower than that for boys (El Tawila 1997: 59–60).

During the 1990s, government and NGOs supported initiatives to increase the proportion of girls in school and to eradicate illiteracy among rural women. They found that, especially in rural areas, many schools were too far from home for girls to attend. As one elderly man said: "We wish all girls, women and men in our village could get an education, but we cannot afford it, nor can we allow our girls to go long distances on their own" (Ibrahim et al. 1999: 91). Moreover, many parents did not think that the schools, especially mixed sex schools, provided an appropriate setting for their daughters' education.

The Ministry of Education recognized the need for innovative approaches to girls' education. Accordingly, in 1992 it established a program of free "girl friendly" community schools. Under the terms of the program, members of the "community" (variously defined) were expected to allocate a schoolroom, and manage the school. The Ministry supplied books and equipment, so that there would be no hidden costs for the children's families. The child-centered approach to learning used games, art work, and stories relevant to children's daily lives; the schools had flexible hours so that children could continue with domestic and farming tasks. By 1999, an estimated 300 community schools existed in rural Egypt, with girls making up 70% of the students.

Although the pupils in community schools worked for the primary leaving certificate (and pass rates were often higher than in conventional public schools) their education was not seen simply as the first step in the process of obtaining a certificate. The objective was far broader. It was to promote a healthy life, in a safe environment, with a level of literacy that empowered girls, and boys, to take advantage of available opportunities. Working to the same end, women's literacy classes were also held in these community schools. Other activities included the creation of water and sanitation programs and income-generating activities (Moehlmann 2001). Several of these programs, especially in women's literacy, were also supported by the National Council for Women. These initiatives are now entering the mainstream, and are no longer seen as experimental. They demonstrate how a community-based approach has the potential for empowering women, and improving education and health.

Looking at tertiary education, at the other end of the educational spectrum, we find that young women are enjoying greater access to university education. In 1995, approximately equal numbers of females and males were enrolled in the faculties of medicine, pharmacy and dentistry (Nassar 1996: 21). This training equips them for professional level employment, especially in the health field, where they can reach other women and children.

Gender and employment

In Egypt, following the Revolution of 1952, formal sector employment was dominated by the public sector in which, by law, women and men have equal access to job opportunities and have comparable salaries. By the mid-1990s, there were more females than males working in this sector: 40% of teachers, 27% of physicians, and around 68% of all nurses were female (Nassar 1996: 20). An increasing number of women physicians were actively involved in disease control activities, and in health promotion. With the revival of training programs for female nurses in the 1960s, women began to replace the male *tomargis* who had worked as hospital assistants. Today, the majority of nurses in the Rural Health Units, the main facilities for the screening and treatment of schistosomiasis, are women.

By the 1990s, as we saw earlier, the employment outlook for the increasing number of high school and university graduates was not as favorable as it had once been. Moreover, employment opportunities overseas—mostly for men—were scarcer than in the previous two decades. Those who did have government jobs were reluctant to give them up, even though the pay was very low. Many people holding such jobs found it necessary to supplement their income with a second job, often in the informal sector.

In many rural Delta communities, the proportion of men and women working full-time in farming has fallen dramatically in the last two decades. Farming as a full-time occupation has been replaced by part-time farming, seen as a way to supplement the low wages earned in government jobs. Children, too, often work in the fields after school or during the long summer holidays. Surveys that only ask rural dwellers for their main occupation overlook the extent of part-time farming for women as well as for men, yet, as we will see, this activity is likely to lead to exposure to canal water and hence to schistosomiasis.

Women as housewives

Most rural women think of themselves primarily as "housewives," *sitt bayt* or *rabit manzil,* literally "women of the house." In the 1996 census, 77% of rural females over the age of 15 were identified by themselves or by their husbands as "housewives," compared to 84% in 1986. Rural women work long hours inside and outside the house. Their unmarried daughters work alongside them, learning their lifetime domestic tasks.

Taken literally, the term "housewife" is misleading, since in fact it encompasses a wide range of productive tasks, such as looking after livestock and chickens, selling produce, and working in the fields. Married women often sell vegetables at small stalls along the narrow streets and alleys of their village, or keep small shops selling inexpensive, everyday items such as patent medicines, matches, and school supplies. In monetary terms, these activities provide only a meager return for the effort involved. In consonance with their role as "housewife," and with religious precepts, women do not hand over the money they earn to their husbands. When they decide how to dispose of it, they generally spend it on their children, rather than on their own personal needs.

Women and water

Women's water-related activities are a central part of their daily round of tasks, and, as we shall see, are likely to result in exposure to schistosomiasis. As safe water and sanitation are not available in all households in rural areas, women are responsible for making a series of choices involving the maintenance of the domestic water system. They must collect and store the water, and, when they have used it, get rid of it. Although larger villages have a piped water supply, not all households have an individual connection. Women living in households without one must bring water into the house from a public tap, or a shallow handpump, or—more rarely—from the canal. A 1996 national sample survey found that just under half, 44%, of rural households had household connections (El Tawila 1997: 68). In the other 56%, women and girls had to fetch water from outside the house.

Any water coming into the house through a tap, or brought in from outside, must be removed in some way from the house. Yet, few Egyptian villages with piped water systems have an effective and safe system to remove sullage (waste domestic water) and sewage. Extensive programs, initiated in the 1960s, provided rural areas with potable water in the form of shared standpipes and, later, individual household connections. Unfortunately, these were not accompanied by drainage or sanitation systems to remove sewage or the wastewater, now produced in far greater quantities than when householders did not have access to taps in their houses (White and White 1986). The reason given for the failure to supply adequate drainage/sewerage for rural communities was that it cost even more to provide than the piped water systems. This has resulted in a serious environmental problem for many rural settlements, as well as extra labor for women.

In households that are not supplied with an adequate drainage or sewerage system, or those that have latrines without a tank that can absorb large quantities of water, pouring domestic wastewater into the latrine will cause flooding, or at the very least, entail the expense of more frequent emptying of the holding tanks. In such a situation, women have no choice but to carry this wastewater out of the house, and throw it in the street or in the canal. Many women solve the problem of disposing of wastewater by carrying out tasks that require a lot of water outside the house, alongside the canal (Watts and El Katsha 1995; El Katsha et al. 1989).

Gender and power

On many levels, there is a difference between what people say about gendered power relations and what actually happens in the family setting. Researchers working in rural Egypt (and elsewhere) have in the past tended to disvalue women's activities. This has happened because women's activities occur in a sphere separate from that of men and often unknown to them. Also, men have taken great pride in telling western male researchers about their dominant position in the family and in local politics. In general, in these situations women have had fewer opportunities to express their views (Singermann 1994; Rogers 1980: ch. 4).

While women's power is rooted in the household, it should not be forgotten that it extends beyond it, to informal networks and collective institutions and activities that sustain daily life. These women's networks, as described in a "popular" district in Cairo, are largely responsible for the families' access to "public goods" such as education, health care and subsidized food (Singermann 1994). The same is true in rural areas. If, indeed, the familial ethos and family relationships are the core of the nation, then it must be recognized that women, as much as men, shape these norms (Joseph 2000: ch. 1).

Without openly flouting the authority of their husbands, women use their roles in child-care and socialization, and in the economic sphere, to forward their own concerns. They run their own revolving credit organizations, and have control over the money they earn. For example, a woman, in defiance of her husband, may spend money on her daughter's schooling because she believes that "the education of a girl is a treasure." Like gold jewelry (a traditional source of emergency cash for women) education gives a young woman a measure of independence in a world nominally ruled by men.

The extent to which women can move freely about the village, and beyond, varies from place to place, and from family to family. Some can move freely about the space of the village and go to nearby settlements, particularly if they do so in the company of female relatives or friends. However, some have little freedom of movement, and are not able to leave the house without their husband's permission, or without a male relative as a chaperone. The most severe restrictions on women's mobility are found in rural Upper Egypt. These restrictions are likely to be most strictly enforced for women whose husbands have little education, for young women immediately before marriage, or in the early years of marriage when they are under the close supervision of their mother-in-law (EDHS 1995: 195–96).

Gender relations are in flux in rural Egyptian settings and, in addition to increases in education and employment opportunities, many other factors are involved in bringing about change. For example, in the late 1970s and early 1980s, the migration of rural young men to Egypt's large cities and to other countries in the Middle East temporarily altered the balance of responsibility in many households. The rapidly expanding metropolis of Cairo, which grew from 2 million in 1947 to 14 million in the mid-1990s, had always been a magnet for men seeking employment. Before the invasion of Kuwait and the Gulf War of 1990–91, men also left Egypt's rural areas to seek employment in Iraq and other countries in the Middle East.

In their husbands' absence, some women took on increased family and economic responsibilities, making decisions that previously would have been made by their husbands. Yet, because of men's legal authority and the strength of (male) opinion, few women could maintain their position when their husbands returned to take over the roles their wives had temporarily filled in their absence (Khattab and al-Daeif 1982).

Rural women were usually aware of the informal authority they had in their community. For example, in the late 1980s, some of our team members, carrying out an action research project in two Delta villages, found that women readily took the initiative in water and sanitation improvements that were directly related to their daily tasks. They had clear ideas about what kinds of interventions were feasible, what would not work, and what they could actually do for themselves. In these Delta villages, local women collected money from other community members to upgrade public taps (standpipes), and kept the area around the improved taps clean. They also made most of the suggestions that were used in the implementation of a garbage collection system. Thus, at the very least, it must be recognized that women's activities outside the narrow confines of the home are of greater significance than is usually thought.

One of the sources of women's power is their awareness of when it is most appropriate to work behind the scenes, persuading their husbands to take action. This happened in the two Delta villages mentioned above. Recognizing that the relevant local administrators were male and would only listen to men, when a standpipe needed to be provided women turned to their husbands, who were better able to deal with the Village Council and district and governorate officials. Thus, the men saw themselves as the main actors initiating the planning and construction of a standpipe and overseeing its mechanical maintenance. Women also had some input into decision making as members of committees dealing with water problems in the village (El Katsha and Watts 1993: ch. 3). Yet, even in this Delta village, some women claimed that they were too busy doing their daily chores to be concerned about things that affected all village women. They said they had limited leisure, and the daily round was more than enough to keep them busy. As one poor women in a hamlet in the governorate of Fayoum (south of Cairo) put it: "We help our men, there is no time for us to rest, we are peasants" (Talaat 2001:31).

Some women in Egypt are unaware of their legal rights, and of the opportunities that may exist for bettering their position. These points are illustrated by the contrasting experience of women in two village development projects that were specifically directed toward women. In one rural credit project, women were encouraged to take loans to begin raising and selling chickens. Although this activity was recognized as being within the women's sphere of responsibility, men insisted on accompanying their wives to the office where the chicks were being distributed, and signing for the down payment (Sholkami 1988: 120). Thus, in practice, the men could control their wives' activities.

In another women's credit program, run by a women's NGO in a rural area of Minya governorate, women managed to maintain control. The leading women participants obtained the identity cards they needed to open a bank account. They admitted that before this time they thought that they did not need an identity card, as their husbands already had one. However, once they obtained an ID they realized that it entitled them to vote in local and national elections. Their husbands could not tell them how to vote as their right to a secret ballot was enshrined in the constitution, whatever their husbands might think.

Villagers need identity cards to interact with almost all government agencies and NGOs, as well as to vote. Yet, as of 1995, in Egypt overall only 38% of women over 18, compared to 95% of men, had IDs. The complex bureaucratic procedures needed to acquire such cards were especially intimidating for illiterate, rural women (Nagi 2001: 265–66). In part for this reason, fewer women than men actually register to vote. In 1986, of 22 million registered voters, only 3.8 million (17%) were women; 5 million women were registered to vote in the 1989 parliamentary elections (Gomaa 1998: 6). Few women have been elected to parliament, though the government regularly makes up for this deficiency by making special appointments of women MPs.

Another way of looking at the participation of women in Egyptian affairs is to reflect on the increasing number of women in influential positions in government, who act as role models for young women. On the ministerial level, the number of women in recent cabinets has fluctuated between two and three, mainly in the area of social affairs, and the recently created environmental post. Elsewhere, women are increasingly being employed in responsible positions in which they have authority over men, and in which they attract media attention. For example, in Munufiya governorate in the mid-1990s, a female physician was in charge of the Ministry of Health Endemic Diseases Unit section, and supervised the Schistosomiasis Control Program. Similarly, in rural health units almost all the nurses, and an increasing number of physicians, were women. Married women physicians and nurses were more likely to live in the village where they worked and to stay longer than their male counterparts, thus providing an element of continuity in a system in which senior staff are frequently transferred.

Similarly, in the schools, an increasing number of teachers are women. As we will see later in the book, these altered patterns of authority, in which gender differences are more evenly balanced than in the past, may encourage a more flexible approach to the challenges brought about by rapid demographic and economic change in rural communities.

In this chapter we have explored ways in which rapid change, especially as it relates to gender roles and behavior, affects schistosomiasis-related issues at the community level. Exposure to infection at canals can be seen in the light of family and community attitudes about appropriate behavior for women who carry out domestic tasks at canal-side sites, and for family members who share farming responsibilities. The decisions of family members about when and where to go for diagnosis and treatment, and what happens when they attend the rural health units, can be understood within the context of gender roles and gender behavior. But before exploring these issues in the local setting, in two communities in Munufiya governorate, we need more information about schistosomiasis, its transmission and impact, and how strategies for control—both prevention and cure—have developed in Egypt. These we provide in the next chapter.

3

Schistosomiasis: A Global Public Health Problem

The global scene

On a global scale schistosomiasis does not attract the attention given to diseases such as malaria or TB, or at the most sensational level, HIV/AIDS. One reason is that schistosomiasis is a chronic, debilitating illness, mostly experienced by poor people in remote rural areas who are ignored by the media. When rural people, especially children, first become infected they suffer intestinal disorders, nutritional deficiencies and debilitation. Life threatening illness can occur when the internal lesions initially caused by the schistosome worm or its eggs develop into cancers or other serious conditions (WHO 1993: 2–4). But schistosomiasis rarely appears as a cause of death in mortality statistics. This is because many of the long-term chronic diseases triggered by the schistosome parasite are attributed, not to schistosomiasis, but to cancers ("malignant neoplasms") and other chronic conditions.

In Egypt, thanks to the introduction in 1988 of a free, nationally available drug treatment for schistosomiasis, praziquantel, the majority of people who become infected are treated and cured and therefore need not suffer from the long-term life-threatening complications that were common before this treatment was available. However, few of the countries south of the Sahara (where about 85% of the global cases are found) are able to afford a nationally based program like that of Egypt. The consequences can be seen by applying a relatively new indicator, Disability Adjusted Life Years (DALYs) that measure the ill-health and death resulting from specific diseases. Although this measure underestimates death due to schistosomiasis (for the reasons mentioned above), it provides a useful alternative view of the impact of the disease in endemic areas (Zou 2001; Barker and Green 1996). Information for 1999 showed that 87% of the global total of DALYs due to schistosomiasis were in Africa south of the Sahara (defined as the WHO Africa region). It is also worth pointing out that, globally, females were found to suffer 39% of all DALYs lost (WHO 2000: 170–71).

Three forms of schistosomiasis are of global public health importance. Although they are similar in many ways, each has its own distinctive characteristics, and geographical distribution. The first is *Schistosoma haematobium*, urinary schistosomiasis. It is found mostly in Egypt, and in sub-Saharan Africa, as shown in figure 3.1. The second is *Schistosoma mansoni,* intestinal schistosomiasis. It has recently expanded its range in Egypt and is widespread in Africa; it is also found in Brazil. The third form, *Schistosoma japonicum* is now found mainly in China. It is the only form in which human reservoir hosts play an important role; hosts include domesticated animals, such as cattle, pigs, and dogs. Two other forms have been identified as locally significant, *S. mekongi*, in the Mekong valley, and *S. intercalatum*, in parts of central Africa.

Figure 3.1: The distribution of *Schistosoma mansoni* and *Schistosoma haematobium* in Africa

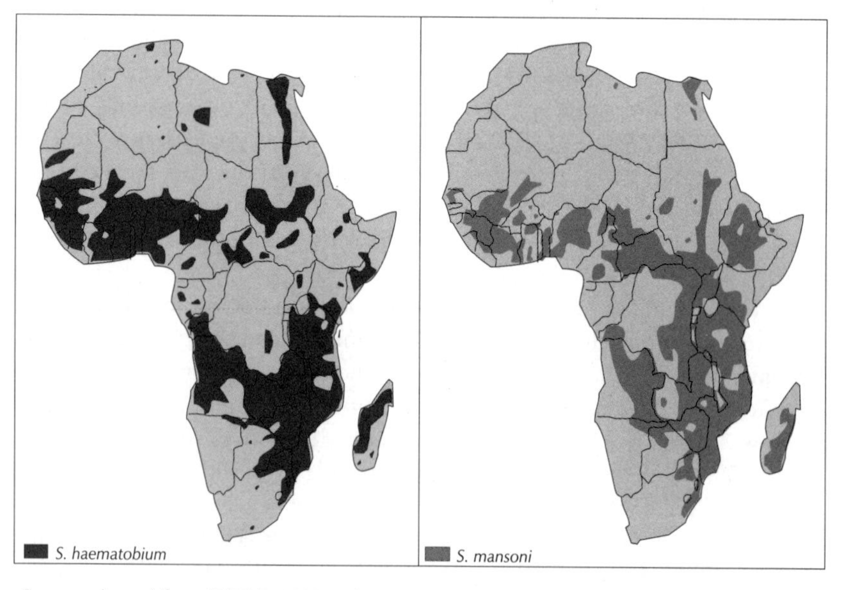

Source: adapted from WHO 1993: 16–17.

The schistosomiasis transmission cycle

Schistosomiasis is transmitted in specific environmental settings, in or near canals, rivers, and lakes containing schistosomes and their intermediate snail hosts. As shown in figure 3.2, two human water related behaviors are involved, exposure and contamination. A water source becomes contaminated when an individual excretes eggs (contained in urine or feces) in or near a watercourse. The eggs hatch as free-swimming *miracidia* (a larval stage of short duration) which must penetrate a suitable snail host. Within the snail host, a single *miracidium* can produce as many as 400 *cercariae* daily (the final larval stage) which are released from the snail into the water. Humans become infected when they enter water inhabited by these *cercariae*. The *cercariae* penetrate the skin and eventually reach the liver where they mature to become male and female worms which pair off and migrate to blood vessels around the gut or bladder. The female worms then produce eggs. The eggs of *S. haematobium* are expelled during urination, and those of *S. mansoni* during defecation (Farley 1991: 8–9).

Figure 3.2: The transmission cycle of *S. haematobium* and *S. mansoni*

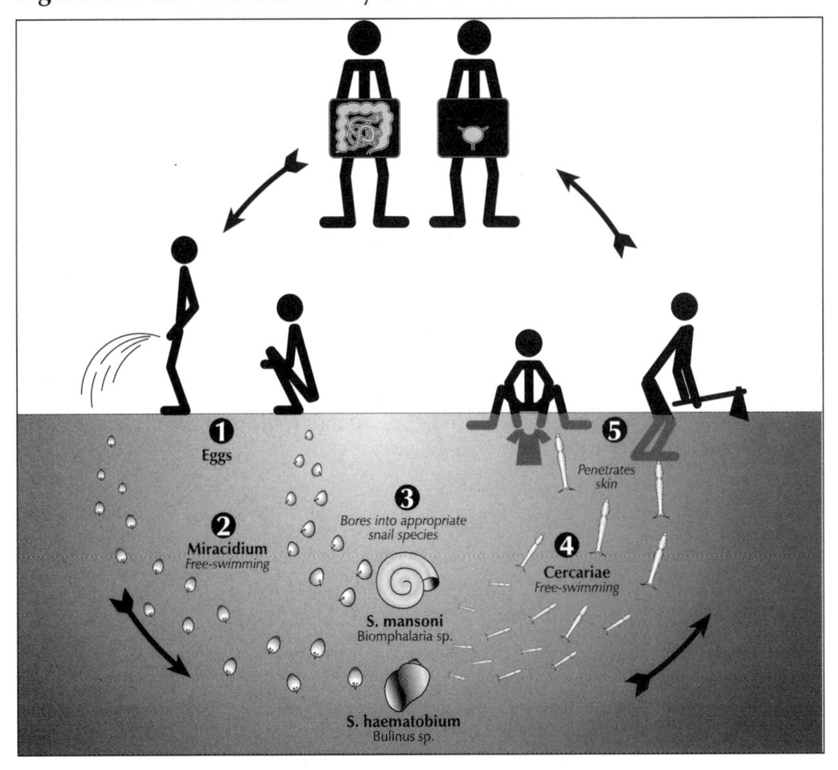

Although the different types of schistosomiasis share many features, in terms of biology they are very distinct parasites. Each has its own distinctive vector snails, survival patterns in the human host, pathology, and path of egg excretion. They also differ in their impact on human morbidity (illness), and in ease of diagnosis and treatment. As we have seen, *S. mansoni* eggs are expelled in feces, and those of *S. haematobium* in urine. As these two types affect large numbers of Egyptians they deserve to be discussed separately.

Schistosoma mansoni

S. mansoni is the predominant type of schistosomiasis in the irrigated farmlands of the Nile delta, the site of the research project described in this book. After the larvae, free swimming *cercariae*, penetrate the skin, the worms mature in the liver and then migrate to the gut mesenteries, the membranes around the intestines, and it is in this organ that they inflict their major damage. When a person is first exposed to *S. mansoni* infection, she or he may experience chills, fever, and diarrhea. Other symptoms include weakness, fatigue, irregular bowel movements, abdominal pain, bloody diarrhea and blood in the stools. However, given that these symptoms are often nonspecific, and can result from other causes, they may not be very useful for diagnosis.

The *S. mansoni* worms can survive in the human host for up to twenty or twenty-five years, all the time producing eggs that are excreted in stools. Thus an infected individual may expel eggs in a canal or another aquatic environment and initiate a cycle of human infection long after public health officials had thought that transmission had ceased locally. The long life of worms in their human hosts and the production of eggs also mean that organ damage can continue long after the initial infection (Editorial 2000).

The long-term impact of untreated *S. mansoni* infections includes liver fibrosis and hepato-megaly (enlargement of the liver). However, these conditions take many years to develop and may remain largely asymptomatic until the disease is well advanced and it may be too late to do much about it. A distinctive symptom of advanced disease, with a high death rate, is hematemisis, internal bleeding. Thus, it is doubly important to identify and treat people with *S. mansoni* as soon as possible after infection.

The infection can be detected by parasitological examination of a stool specimen. If found positive, treatment can be provided; this is now considered effective for about 60–80% of *S. mansoni* infections, depending on circumstances (El Khoby et al. 2001). As we shall see, the collection and testing of stool specimens at health facilities presents gender related and logistical problems quite different from those involved in the collection of urine samples for the identification of *S. haematobium*. Also, the prevention of the contamination of water bodies that results from indiscriminate defecation involves considerations which are somewhat different from those involved in contamination due to urination.

Schistosoma haematobium

S. haematobium worms mature in the liver and migrate to the blood vessels around the bladder, where most tissue damage occurs. Eggs are then expelled in the urine. The major symptom is hematuria, blood in the urine, which occurs when the spiny eggs puncture the walls of the urinary bladder. This has given rise to descriptions of "menstruating men." In some endemic areas hematuria was so common that it was regarded as a rite of passage among adolescent males, who usually comprise the most highly infected sector of the population (Akogun 1991).

Adult *S. haematobium* worms have a life span in the human body limited to five years. Thus, serious long-term damage is likely to occur mainly in cases of repeated infection that remain untreated. In addition, the shorter life span of *S. haematobium* (compared to that of *S. mansoni*) affects the dynamics of transmission, with individuals excreting eggs into the aquatic environment over a shorter period of time.

In general, on initial infection, between 50% and 75% of individuals experience hematuria and problems associated with frequency and pain on urination (WHO 1993: 41). In both *S. haematobium* and *S. mansoni*, worms and especially eggs can migrate to many parts of the body and cause lesions. *S. haematobium* has, thus far, been more commonly associated with lesions of the female and male genital organs than has *S. mansoni* (Feldmeier et al. 1995; Feldmeier et al. 1999).

Given that *S. haematobium* eggs are expelled in urine, a parasitological diagnosis can be made from a urine sample. For both schistosomiasis types, ultrasonography, which provides an "X-ray" picture of soft tissue, and palpation can detect the extent of tissue damage. In the case of *S. mansoni*, ultrasonography can identify the enlargement of the liver and spleen, and the resulting portal hypertension that occurs when the blood flow to the liver is blocked.

The impact of schistosomiasis on daily life

We have little detailed information about the extent of actual illness, morbidity, and the ways in which schistosomiasis can affect the daily life of those infected. In the late 1960s and 1970s, reflecting a policy concern with the "cost effectiveness" of health interventions, researchers assessed the work capacity of male cane cutters on sugar estates in Sudan and Tanzania. They found small differences in measures of labor efficiency between the workers who were infected with *S. mansoni* and those who were not, but both studies suggested that it was difficult to attribute such differences in productivity solely to the infection (Fenwick and Figenschou 1972: Collins et al. 1976). Other researchers found that, in Sudan, males aged 18–45 who were excreting a large number of *S. mansoni* eggs were less able to perform heavy physical tasks than those who were uninfected (El Karim et al. 1980). These studies examined men who were relatively fit and able to perform their daily work. As researchers excluded those who were too sick to go to work, they clearly underestimated the impact of the disease at the community level.

The impact of schistosomiasis on women has been neglected partly because they were seen as being simply "housewives," far removed from what male interviewers considered to be the economically productive sphere; such work supposedly had no monetary value. However, even if women do not engage in wage labor (that could conveniently be given a monetary value) many women in low income countries do make a monetary contribution to the household, working in the informal sector as small traders and farmers. Aware of this research bias, in the late 1980s Melissa Parker conducted a detailed anthropological study of the impact of *S. mansoni* on young women in the prime of life in the cotton growing area of the Gezira, Sudan. She concluded that, although women with *S. mansoni* spent less time in the fields, they managed to harvest more cotton in that shorter period. She suggested that this was because women who were unwell felt that they must overcompensate for their weakness and make a special effort when picking cotton, a high value crop which contributed significantly to the family income. However, during the rest of the day these women spent less time on other agricultural tasks, on domestic tasks, and on personal care. Parker concluded that "the impact of *S. mansoni* infection on daily activities is complex and variable" (Parker 1992: 888; see also Parker 1993).

Targeting school-age children

Biomedical researchers and health planners are agreed that school-age children suffer disproportionately from schistosomiasis and that in many endemic areas, infection rates are higher for children than they are for adults in the prime of life. On average, children excrete more eggs than do infected adults. As they tend to be rather careless about personal sanitary matters, in that they will casually shed urine or fecal matter while swimming or larking about in the water, they have a disproportionately larger role in transmitting the infection than do other age groups. Studies have also shown that, after treatment, children are more often reinfected than are adults. This may be due to their higher levels of water contact, and to immunological factors that provide adults with resistance to reinfection. It has been suggested that resistance is not due solely to exposure to infection but may also be due to age itself (Gryseels 1994).

Children also often suffer more acutely from the early stages of the infection, partly because they harbor a large number of worms that produce many eggs, and partly for immunological reasons (as suggested above). Children who remain untreated are likely to go on to experience long term, potentially fatal, conditions later in life.

Early stage schistosomiasis infection among children has been associated with abdominal discomfort, diarrhea, and lassitude, as well as anemia and malnutrition, as parasites compete with their young human hosts for the available food supply. In the case of *Schistosoma japonicum* (not found in Egypt) infected children suffer delayed physical and mental development. A recent study of Chinese primary school children found an improvement in cognitive functioning following treatment (Nokes et al. 1999). In Egypt, among children infected with *S. mansoni* and *S. haematobium,* the evidence for developmental problems is less clear cut. However, a small recent study of 80 children in a primary school in Kafr al-Sheikh governorate in the northern Nile delta showed that children infected with *S. mansoni* performed significantly less well on comprehension, vocabulary, and picture completion sections of the IQ tests than uninfected children (El Morshedy et al. 1999).

In Egypt, school-based control strategies have long been part of the schistosomiasis program. However, their success depends on the proportion of children who attend school. Here, as in most low income countries, fewer girls than boys attend school. Later in the book, we will describe the development of a gender sensitive school-based *S. mansoni* screening and treatment program for school children, especially girls.

Since the late 1980s, concern for the health of school-age children has been given added visibility by a series of programs supported by UNICEF, the World Bank, and by the Partnership for Child Development (founded in 1992). Planners and policy makers recognized the great improvement in the health of children under the age of five that had occurred since the late 1970s, partly due to Child Survival programs that promoted interventions such as oral rehydration therapy (in cases of diarrhea) and vaccination. They saw the need to go on to develop effective strategies to improve the health of older children. Chief among these was the treatment of parasitic infections. Many children in Egypt and other low income countries suffer from concurrent parasitic infections, including schistosomiasis, which result from exposure to a contaminated environment. In addition to causing abdominal discomfort and other symptoms, parasitic infections hinder the absorption of nutrients, thus contributing to malnutrition and anemia (Bundy and Guyatt 1996; El Sayed et al. 1998).

Diagnosis for schistosomiasis in the schools, particularly the use of stool samples for *S. mansoni*, can at the same time identify the eggs of a number of other common parasites. These include oxyuriasis, (otherwise known as pin worm or thread worm) and ascariasis, both transmitted by the fecal-oral route, and hookworm, transmitted from feces to the soles of bare feet. Recent studies in Ghana and Tanzania have found school-based diagnosis and treatment for common childhood parasitic infections, including schistosomiasis, to be cost effective, which is to say that it is cheaper than other strategies such as mass treatment or screening at a health unit (Partnership for Child Development 1999).

Human behavior and schistosomiasis transmission

As we have already seen, two distinctive human behaviors are associated with the transmission of schistosomiasis: exposure to infection at a water source, and contamination of this water source. The understanding of these culturally and socially patterned behaviors, the contexts within which they occur, and the reasons people give for behaving in this way provide important insights into the control of the disease (Huang and Manderson 1992).

To date, most studies of the behavior of people at canals, rivers, and lakes where they may come into contact with the schistosomes have not gone beyond looking for information that can be quantified. Epidemiologists have either asked respondents about their water contact activities or have observed such activities directly and have presented their findings in a numerical form. There have been few follow-up explorations to find out *why* such water contact behaviors persisted. As a recent paper by medical anthropologists has pointed out, there is a real need to understand the underlying "cultural logic" of this behavior (Brown et al.1996: 207–8). From the point of view of our research, we also needed to develop ideas for the behavior change that is necessary to avoid repeated exposure and infection. As we will explain later, behavior change usually involves two parties, the person actually in contact with the water, and the agencies responsible for the building and maintenance of canals, domestic water, and drainage systems. Such investigations formed an important part of our research in the Nile delta, and are described later in the book.

Identifying and understanding the behavior associated with the contamination of water sources is more difficult than studying exposure. An individual's excretory activities, in rural Egypt as in most (though not all) areas, are private acts that can rarely be observed, and are not generally discussed with strangers. In addition, beliefs and behaviors associated with defecation practices (in the case of *S.mansoni*) are likely to be different from those associated with urination and the transmission of *S. haematobium*. The only observational study of defecation practices in and around water courses took place in the Gezira irrigated area of Sudan. It involved the use of jeep traverses and field glasses to identify excretory acts likely to transmit *S. mansoni* (Cheesmond and Fenwick 1981). A study such as this is extremely valuable, but is difficult to replicate due to its intrusive nature and the fact that almost all people over the age of four consider defecation to be a private act. However, as a single excretory act can contaminate a canal near an area of human activity, an understanding of such acts remains crucial to attempts to control disease transmission. In addition to individual acts, there is also some evidence from our Nile delta study that latrine effluent containing schistosome eggs (among other pathogens) has been illegally dumped into canals.

Schistosomiasis is highly focal. It occurs only at sites where humans are exposed to, and contaminate, a water source containing vector snails. Some villages have no cases, while in others nearby, half the population is infected. Yet, as we shall see later, few attempts have been made to find out why these great differences exist.

The development of water resources and schistosomiasis transmission

Globally, the damming of rivers for irrigation and hydroelectric power has been responsible for an increase in schistosomiasis infection. The artificial lakes and slowly flowing irrigation canals brought into being by the dams create new opportunities for employment and the in-migration of people who are not aware of the dangers found in schistosomiasis-riddled waters. Dams in Africa, such as the Volta River Project in Ghana, the Kariba dam in Zambia, and the Kainji Dam on the River Niger in Nigeria, have all been associated with a massive extension of schistosomiasis (Hunter et al. 1993).

One of the most notorious examples occurred in the Senegal River delta. Here, in the late 1980s, a series of dams to regulate the water flow for irrigation opened up new areas for human settlement. But the new canal network also provided a marvelous environment for *Biomphalaria pfeifferi*, the local vector snail of *S.mansoni*. Compared to other snail vectors, *Biomphalaria pfeifferi* has an unusually long life span and expels an unusually large number of *cercariae*, all looking around for a human host. This resulted in an explosive epidemic among the new settlers. Alarm bells rang in the scientific community when it was revealed that the cure rate for *S. mansoni* locally was as low as 18%. It is now thought that the specific local conditions were responsible for low cure rates: continuous intense transmission, patients' high worm loads, and rapid reinfection (Gryseels et al. 2001; Ernould et al. 1999). As we will discuss in the next chapter, Egypt, where 95% of agricultural production depends not on rainfall but on irrigation water drawn from the River Nile, presents an excellent example of the complex interrelationship between human behavior, irrigation, and schistosomiasis.

Human behavior, gender, and biomedicine

Having identified the main features of schistosomiasis, drawing mainly on the findings of researchers in biomedicine and the biomedical subdiscipline of epidemiology, we will briefly explore the relationship between biomedicine and social research, especially as it relates to schistosomiasis. As social scientists working alongside medical scientists and epidemiologists, we recognize that biomedical scientists have long neglected human behavior and gender issues. In spite of a seminal paper in 1979 by Frederick Dunn on the behavioral aspects of parasitic diseases, it was not until around 1990 that biomedical and epidemiological researchers began to consider seriously the importance of human behavior and gender, and how they related to disease prevention. The first stage in this process was to encourage research collaboration between the biomedical and social sciences, or more specifically between the subdiscipline of tropical medicine and the social sciences.

Attempts to foster this collaboration began with the establishment of TDR (Tropical Disease Research—officially the UNDP/World Bank/WHO Special Programme for Research and Training in Tropical Diseases) in 1974. TDR was established with the specific mandate to strengthen institutional capacity in tropical disease research, including schistosomiasis, in the countries in which the diseases occurred. In addition to biomedical research, TDR supported social, behavioral and economic research. It encouraged an active input from social scientists, providing training and an intellectual and institutional environment favorable to the collaboration between biomedical and social scientists (Morel 2000).

In 1989–90, TDR, in collaboration with the Canadian International Development Research Center (IDRC), organized a competition for papers in the neglected area of women and tropical diseases. Of the ten papers selected for publication, three focused on schistosomiasis. Hermann Feldmeier and Ingela Krantz presented an inventory of needs for research on women and schistosomiasis (Wijeyaratne et al. 1992, 100–33; Feldmeier and Krantz

1993). Michelson (1993) reviewed the literature to identify "predisposing factors" that affected the transmission, prevalence (level of infection), intensity (measured in terms of egg output of those infected), and morbidity from schistosomiasis. He identified infections of the genital system as a particularly important area for future inquiry. Only one paper, by Melissa Parker (quoted above), presented the findings of actual field research, on the impact of schistosomiasis on women in rural Sudan.

After reviewing the papers, the organizers of the IDRC competition highlighted the lack of research on women and gender issues, and of collaboration between social and biomedical researchers (Wijeyaratne et al. 1992). In retrospect, the competition also reflected the continued concern with women (and with the biomedical focus on physiological differences between the sexes) rather than the broader concerns of gender, defined as the interactions between men and women.

When seen from the perspective of the social sciences, the awareness of gender issues in biomedicine is sometimes more apparent by its absence than by its presence. For example, a major task for epidemiologists (who have conducted much of the field work on schistosomiasis) is the identification of "risk factors" that appear to be significantly associated with schistosomiasis. As schistosomiasis is directly related to exposure to certain water sources containing vector snails and schistosomes, the relationship between "risk factors" and infection would appear to be relatively unproblematic. Epidemiologists found that males were usually significantly more infected than females; the reverse was very rarely true. Thus, they identified being male as a risk factor for schistosomiasis.

This, however, is not the way a medical anthropologist looks at the issue. For someone in that discipline, "risk" is a result of social or political forces which predispose certain groups of people to infection. In contrast, for most epidemiologists the relationship (expressed as a statistical correlation) between infection status and characteristics such as sex, age, occupation, or place of residence is the result of the cumulative effects of individual behavior. They rarely probe further into these relationships (Brown, Inhorn and Smith 1996: 197, 348–49).

An integrated approach to schistosomiasis control

The balanced understanding of schistosomiasis transmission and treatment requires an interdisciplinary as well as a multisectoral approach. The use of praziquantel (as in Egypt since 1988) can contribute to a sharp decline in the number of infections, and can control morbidity due to the disease. This is primarily a biomedically oriented, curative approach to the control of schistosomiasis. However, since the discovery of the disease transmission cycle in 1915, public health specialists have recognized the importance of prevention. This involves developing strategies to change behavior among policy-making officials as well as among "at risk" local people. Prevention also means destroying the snail vectors and changing the ecology of canals, as well as upgrading the water supplies and sanitation facilities.

This multi-faceted approach to schistosomiasis control has come to be known by international health specialists as the "integrated" approach. As defined by the World Health Organization in the period leading up to the introduction of praziquantel, the essential elements of such a program were diagnosis and treatment, snail control, water and sanitation improvements, health education, and "community participation" (local involvement in control strategies). WHO also noted that these programs needed to be supported by a system of record keeping and reporting to assist health staff at various levels to provide services when and where they were needed—in epidemiological and public health terminology this is known as "surveillance."

The balance between the various elements in a control program, and how they are applied, depends on local conditions in endemic areas and the resources of the countries that develop such a policy (WHO 1993: 2, 8). However, not all endemic countries have the resources available to Egypt to tackle this insidious infection. Many African countries south of the Sahara, where schistosomiasis also remains a serious rural health problem, cannot afford to purchase praziquantel or to maintain an effective national network of primary health facilities.

The disparate elements of a schistosomiasis control program are "integrated" in the sense that they are usually coordinated on the national level by a Ministry of Health. However, it is important to recognize that they utilize skills from various disciplines that, in turn, incorporate rather different disciplinary assumptions about how the practitioners should go about their tasks. For example, diagnosis and treatment, usually through primary health services, is planned by medically trained experts and carried out by trained physicians, nurses and technicians. Vector control—chiefly the use of chemicals to destroy vector snails—is usually researched, planned, and carried out by biologists. Similarly, engineers of various kinds are needed for programs to improve water and sanitation, and canal and irrigation management. Health education is usually the task of trained health educators.

Since the introduction of praziquantel, treatment has assumed an even more dominant role in integrated programs. Public health specialists have debated the relative merits of "selective" treatment (provided only for those who have been found to be positive after diagnosis) and "mass" treatment (treatment of everyone, or certain groups such as school children, in highly endemic areas, without prior diagnosis). This debate has tended to sideline the important question of the relative role of the other approaches. In the next chapter we examine the genesis of these various approaches in Egypt.

4

The Growing Awareness of Schistosomiasis in Egypt: to 1988

Introduction

Reflecting both the accidents of geography and of the country's exceptionally long history, since the early 1900s more has been written about Egypt's battle with schistosomiasis than has been written about the disease in any other part of the world. This also reflects the fact that the country has been the center of important research into the identification of the disease causal agent, this agent's two hosts (snails and humans) and various strategies for its control.

In this chapter we will deal with these issues until the year 1988, when the Egyptian Ministry of Health made the new drug praziquantel available free for everyone found to be infected. We will begin with a brief discussion of Egypt's changing patterns of irrigation from Pharaonic times onward.

Schistosomiasis and the changing Egyptian irrigation system

Egypt is one of the few countries of the world that depends almost entirely on a single river, the Nile, to provide it with water for cultivation, human consumption, bathing, domestic and industrial activities. Only in a small area on the Mediterranean coast, to the west of Alexandria, is there enough annual rainfall to make cultivation possible without irrigation.

Going back to pre-dynastic times (3,400 BCE), the earliest method of irrigation in the Nile valley and Delta was *basin irrigation*. During the summer inundation, when the Nile was at its widest and deepest, water flowed into low-lying basins bordered by natural levees. Water stored in these basins supplied the crops planted as the water receded. Over time, canals were built that regulated the flow of water into the basins. The first lifting device, the *shaduf*, could only provide water for small plots next to the water course, but the introduction of the waterwheel, in Ptolemaic times (after 332 BCE), extended the area that could be cultivated.

Using this system, most areas of Egypt were limited to a single crop a year, though two crops might be grown on arable land that bordered the Nile itself. These limitations notwithstanding, during the late Ptolemaic period, Egypt exported huge quantities of grain. It was in order to win control of this bread-basket of the eastern Mediterranean that Rome conquered the country in 30 BCE.

From the time of the earliest farming settlements in the Nile valley, around 7,000 years ago (and until the construction of the Aswan High Dam, completed in 1964), farmers benefited from the annual inundation and its gift of fertile silt from further south. Whether or not these early farmers suffered to any extent from schistosomiasis is not at present clear. A small number of mummies of elite people from the New Kingdom period (1567–1085 BCE), who had fished and hunted in swamplands bordering the Nile or in Fayoum, have been found to contain the calcified ova of *S. haematobium*. However, the recreations and lifestyle of these people were not typical of the population of Egypt as a whole. Recently, a number of human remains from the pre-dynastic period, 5,000 years ago, have been found to contain the schistosome circulating anodic antigen, indicating an active infection at the time of death. However, we have no idea of how representative those people were of the overall population in their district, or of the Nile valley as a whole. As we now understand it, schistosomiasis can be highly focal. In any case, although a large collection of ancient Egypt's surviving medical papyri have been deciphered and closely studied, they contain no mention of a disease condition that can securely be identified as schistosomiasis (Nunn and Tapp 2000).

In the twelfth and thirteenth centuries of our era, Arab physicians mentioned hematuria among farmers in the Nile valley and Delta. Larrey, a French physician accompanying Napoleon's army to Egypt in 1798, described hematuria among their own troops (Farley 1991: 44). However, such anecdotal information is not a very reliable guide to the actual level of infection among the population, and it appears likely that schistosomiasis did not become widespread in Egypt until the mid-nineteenth century and the coming of major changes in irrigation practices. Beginning in the time of

Muhammad 'Ali Pasha (1805–48), basin irrigation began to be replaced by what is called *perennial irrigation*, providing water for cultivation all year round.

As part of his project of "modernizing" Egypt and bringing it more fully into the global economy dominated by Western Europe, Muhammad 'Ali began to build barrages at al-Qanatir, 40 km north of Cairo, to control the water supply. The project was completed by British engineers brought in from India after the British conquest of Egypt in 1882. To further stabilize the water sources used for agriculture (largely for the production of export crops such as cotton) the British built the first Aswan Dam in 1902, and gave it added height in 1910.

Using these huge, dammed water supplies (behind the barrages and the Aswan Dam) landowners and government authorities introduced perennial irrigation on a large scale. This meant the building of hundreds of miles of canals in complex networks consisting of main canals, secondary canals and field canals. This new form of irrigation permitted two or three crops to be grown each year instead of only one, as in the past.

As far as schistosomiasis was concerned, it meant that the canals in which the parasites' snail hosts lived flowed slowly and at a constant level for most of the year. They were closed only for 6 to 7 weeks every winter for cleaning. This period gradually became shorter after the construction of the Aswan High Dam that began to fill Lake Nasser in 1964. These conditions proved to be far more encouraging for the survival of the snail hosts of schistosomiasis than the more rapid water flow and the long winter drying-out period typical of canals under the basin irrigation system (Abdel Wahab 1982: 62–64).

A growing knowledge base

We owe our first solid knowledge about schistosomiasis to a medical scientist who came to Egypt in 1850. Theodor Bilharz had received scientific and microscopy training at the University of Tübingen in Germany, specializing in the new field of helminthology (the study of worms). He conducted research at the Qasr al-'Aini Medical School in Cairo (the school and hospital founded by Muhammad 'Ali in the 1820s). Using a microscope to study a patient's urine, Bilharz identified the eggs of the schistosome species that came to be known as *Schistosoma haematobium*, and watched as the embryo emerged from the egg. He suggested that the eggs were related to the worms found, during an autopsy, in the portal vein (which leads from the bowel to the liver) and that they were also related to hematuria (Farley 1991: 50–54).

As the founder of schistosomiasis studies, Bilharz is memorialized in the medical terms first used to describe the infection, bilharziosis, bilharziasis and bilharzia, the last term being still widely used today. The Egyptian people remember him on a daily basis when they refer to the infection as *bilharzia* or *harzi*. The medical research center in Imbaba, Giza, is known as the Theodor Bilharz Institute.

In 1907, Luigi Sambon, an Italian researcher, suggested that, in addition to the schistosome species identified by Bilharz, there was a second species in Egypt. He dubbed this new species *S. mansoni*, in honor of the leading expert in tropical medicine in Britain, Patrick Manson, who was head of the London School of Tropical Medicine from 1899 to 1912 (Farley 1991: 55–59). In the same year (1907) that Sambon identified the second species of schistosome, the expatriate dean and chief surgeon of Qasr al-'Aini Hospital, Frank Cole Madden, published the first book length study of schistosomiasis. Illustrated with several stomach-churning photographs of Egyptian males and females whose genital parts, spleens, livers, bladders and rectums were swollen and dysfunctional, Madden's book raised consciousness of the disease.

The next year, Bonté Sheldon Elgood (1908), an Assistant Medical Officer in the Education Department, in what was perhaps the first professional paper on schistosomiasis by a woman, examined and tested patients at Qasr al-'Aini Hospital and children at schools in Cairo in an attempt to find out how they acquired the disease. She asked them what kind of water they used at home and if they ever came into contact with canals or ponds. These responses, when compared to the individual's infective status, failed to provide clear evidence for the origin of the infection; she considered that, on balance, infection was more likely to be caused by drinking impure water than by bathing in an infected pond or canal. It was clear that the understanding of how human beings became infected—by drinking water or by actually coming into skin contact with the water—was a key to effective control strategies.

In 1914, a young Scottish helminthologist, Robert Leiper, arrived in China to investigate the life cycle of the schistosome on behalf of the British Admiralty, which was concerned about the spread of the infection among their sailors on the Yangtze River. After having reached an impasse in his own research, Leiper heard about the discovery, in China, of the snail intermediate host of *S. japonicum* by a Japanese scientist, Keinosuka Miyairi, who had trained in Germany. Leiper was able to follow it up, and in 1914, he was sent to Egypt by the War Office in London, this time to study the life cycle of the local schistosomes which were thought to represent a threat to the British troops guarding the Suez Canal.

Within a year of arriving in Egypt, Leiper discovered that the two types of the disease there, *S. haematobium* and *S. mansoni*, followed a life cycle that was in most respects identical to that Miyairi had identified for *S. japonicum* in irrigated areas of China. Leiper published his findings in several installments between 1915 and 1918 (Farley 1991: 64–67; Jordan 2000). His discoveries lifted a great load from the minds of officials in the War Office in London, as his work demonstrated that schistosomiasis posed no threat to the occupying British soldiers in Egypt, provided that they had access to safe water and sanitation and did not come into contact with infected canal water.

However, in the years after 1918, when Egyptian nationalist agitation was seriously undermining the foundations of their moral authority in Egypt, British authorities were only mildly concerned about tackling the problem of schistosomiasis among rural Egyptians. For one thing, it was manifestly more difficult to control the disease among Egyptian farmers, who were in constant contact with the canals, the snail vectors, and the schistosomes, than it was among garrison soldiers.

The possibility of controlling schistosomiasis in Egypt was first considered following the discovery, in Sudan, that the drug tartar emetic (antimony tartrate) was effective against schistosomiasis. Dr. J. B. Christopherson, the head of the Khartoum hospital, began to experiment with tartar emetic, a non-disease specific, all-purpose purging poison that was intended to kill off parasites within the human system. Once something resembling a satisfactory dosage was established (one that did not kill the patient), tartar emetic was tested on various village populations in Egypt. Follow-up showed that, at best, around 65% of the people who took the full course of treatment were cured. The full course of treatment required 12 injections over a four week period, each of which, from the villagers' point of view, meant one day in which he or she could not work, earning subsistence pay (or less). Worse still, the drug (well known as a poison) had serious side effects, which deterred patients from completing the course and thus having a 6 to 7 in 10 chance of achieving a cure.

By the early 1920s, the Egyptian Public Health Department, which was coming increasingly under Egyptian control as the British handed over certain government departments to locally staffed authorities, established tartar emetic treatment camps in many areas of the Nile valley and the Delta. At the same time, a parallel program was developed to destroy the snails in canals, using another poison, copper sulfate (Farley 1991: ch. 6).

The next stage in the development of the knowledge base about schistosomiasis in Egypt was associated with the rapidly developing science of epidemiology, a statistically based branch of biomedicine aimed at establishing the extent of the infection, its severity, and the distribution of the different types of schistosomiasis. A key early study, carried out in 1935, based in part on community studies and in part on an analysis of the official treatment records, was that of Dr. J. Allen Scott (1937). Scott, an American epidemiologist, was sent to Egypt under the auspices of the Rockefeller Foundation of New York to assess the status of schistosomiasis in the country and to explore the relationship between infection and perennial as opposed to basin irrigation. Scott found overall high rates of *S. haematobium* in the Nile valley south of Cairo and throughout the Delta. He found *S. mansoni* was restricted to the Delta; although the pattern of infection was patchy, rates were generally higher in the northern parts of the Delta than they were further south. In the southern part of the Nile valley, levels of infection with *S. haematobium* were found to be somewhat higher in areas of perennial irrigation than they were in areas where there was still basin irrigation.

From the early 1930s, Egyptian epidemiologists mapped out what they found to be increasingly high schistosomiasis infection rates in areas of perennial irrigation. They contrasted these with the comparatively low rates of infection in areas of basin irrigation (Khalil and Azim 1935). The unspoken finding here was that the capitalist strategies based on the cultivation of export crops by using year-round irrigation were incompatible with the health of the Egyptian farmer. The Rockefeller Foundation preferred to take a different view of parasitic infection, seeing it as a cause of worker inefficiency. The Foundation first turned its attention to hookworm in the south of the United States. In Egypt in the 1930s and 1940s, it funded a number of studies by expatriate staff which were intended to demonstrate that the introduction of safe water supplies and latrines in rural Egypt would contribute to the control of schistosomiasis transmission and hence increase worker efficiency. For various reasons, the projects failed to yield the desired results (Farley 1991: chs. 6 and 11).

In the 1960s, the Egypt-49 project was established, as a collaborative effort of the Egyptian government, WHO, and UNICEF to test the effectiveness of the chemical Bayer 73 against vector snails. The Egyptian scientists conducting this project, in villages south of Alexandria, were the first to use a new statistical technique, the measurement of "incidence" to assess the effectiveness of a disease control program. They began with a prevalence survey, followed, at a suitable interval after the beginning of the control program, by another survey to identify the new infections, thus providing the measure of "incidence" (Farley 1991: 279). Rather confusingly, Scott's use of the term "incidence" in his 1937 paper had referred to what epidemiologists now call "prevalence." Mohamed Farooq, the director of the Egypt-49 program, also pioneered the observation of the behavior of rural people at canals. His researchers found that females were responsible for 68% of the

2248 activities they observed, suggesting that within the settled areas, domestic activities could be responsible for women becoming infected (Farooq and Mallah 1966).

Following Farooq's study, other researchers published accounts in Egypt and in Nigeria, Ghana, Kenya, Zambia, South Africa, Brazil, and in the West Indian island of St. Lucia, illustrating the great variety of water contact activities implicated in schistosomiasis transmission. They identified the gender and age of those involved, the actual time spent in the water, and the areas of the body exposed. Working in Upper Egypt in the mid-1980s, Helmut Kloos observed adolescent boys swimming in canals and related this directly to infection (Kloos et al. 1990, 1983; see also Brown, Inhorn and Smith 1996: 206–7).

By the late 1970s, epidemiologists were beginning to recognize that an important change in the distribution of the two types of schistosomiasis was taking place in the Nile delta (El Alamy and Cline 1977; Abdel-Wahab et al. 1979). In 1983, a study in some of the Delta villages studied by Scott nearly fifty years earlier found high rates of *S. mansoni* throughout the Delta, with up to 70% infected in some villages in the north. The greatest percentage increases in *S. mansoni* were recorded in the southern Delta, especially in Munufiya governorate, suggesting that this type of infection was moving south. Concurrently, rates of infection with *S. haematobium* had diminished, and it was rare to find a community with a prevalence rate of more than 10%. At the same time, the population of *Bulinus truncatus*, the *S. haematobium* vector snail, had decreased markedly throughout the Delta (Cline et al. 1989). As we shall see in the next chapter, by the early 1990s, *S. mansoni* had become far and away the predominant schistosome species in the Nile delta. This change had important implications for the treatment strategies in the Delta that will be discussed later in this book.

The burden of infection

It had long been perceived that rural Egyptians were more likely to be troubled by schistosomiasis than were the small minority of people who lived in cities. However, in the early years of the twentieth century, given the political realities of the time (Egypt's role in the British managed global economy as a producer of raw materials for export, using intensive irrigation) there was little attempt to assess the burden of the disease in Egypt as a whole. Then, in 1949, Dr. Mohammed Khalil, the official in the Ministry of Public Health responsible for the control of schistosomiasis (and later Minister of Health), and a committed Egyptian nationalist, made an impassioned statement about the impact of the disease on the country. He wrote that:

> Bilharziasis is the greatest obstacle to the progress and prosperity of the Egyptian Nation. 14 million out of its 20 million population suffer from the disease. The disease retards the mental and physical development of the individuals. It lowers the output of the manual labour to the extent that Egypt loses on this item alone £ 80,000,000 a year. The disease undermines the efficiency of the Egyptian Army. The disease is responsible directly for 25% of the deaths in Egypt. (1949: 852).

Khalil based his assessments on a generalized estimate of causes of death revealed in post-mortem examinations at Qasr al-'Aini Hospital in Cairo and on the proportion of army recruits rejected because of stunted growth (21% of all recruits in the northern Delta), which he considered to be due to schistosomiasis (Khalil 1949: 818–19).

Khalil's estimate of the number of Egyptians affected was greater than that of Scott in 1935, who had concluded that almost half of the 15 million Egyptians were infected. However, writing in 1978, F. DeWolfe Miller considered that Scott's figure was an underestimate and that the number infected was probably closer to 60–80% of the total population (Miller 1978: 27). Miller based his estimate on the fact that the screening technique available in Scott's time had been less sensitive than that available forty years later, and thus less able to identify eggs in the specimens of people with light infections. In 1941, 'Abd al-Wahid al-Wakil (later Minister of Health), claimed that 75% of Egyptians were infected with schistosomiasis, compared to 90% with trachoma (a potentially blinding eye disease) and 50% with hookworm (Gallagher 1993: 15). In 1993, the Ministry of Health estimated that half of the rural population (i.e. around a quarter of the total population) was infected with schistosomiasis (El Khoby et al. 1998). Whatever the precise figures, it is clear that schistosomiasis was undermining the health of a huge number of Egyptians. It was also clear that this disease threat could be brought under control.

Schistosomiasis control programs in Egypt

Control programs in Egypt began in the 1920s, using Christopherson's tartar emetic treatment, and copper sulfate for snail control. Faced with expanding numbers of people infected, by the 1960s regional control programs were introduced, using a new generation of drugs, and, in some cases, new molluscicides to kill the snail vectors. In 1969 the Ministry of Health initiated a program to eradicate schistosomiasis in the Fayoum, an intensively irrigated basin linked to the Nile by the Bahr Yusif canal. Treatment was provided with a new drug, ambilhar, and canals were blanketed with niclosamide. As a result, in the short term, the number of people with *S. haematobium* declined markedly. But, by 1985, the level of infections in Fayoum was back to the levels found before the intervention.

In 1977, a control program was introduced in Middle Egypt (covering Beni Suef, Minya and northern Asyut governorates) using the drug metrifonate to treat people, and niclosamide to control snails. Three years later, the project was expanded south, to southern Asyut, Sohag, Qena and Aswan governorates, and eventually covered 12 million people on over two million feddans of land. In 1984, the program was extended to Giza governorate, just south of Cairo. Overall, these government sponsored interventions led to a marked decline in *S. haematobium* infection in the Nile valley, to below 10% in all governorates except for Fayoum (El Khoby et al. 1998; El Khoby et al. 1991).

Also in 1977, the Ministry of Health introduced a new national-level control program which included diagnosis and treatment (including a special program for school children), and molluscicidng. This program coordinated activities in existing regional programs and also included the Nile delta. Although tartar emetic was gradually phased out, it was still in use in the mid-1970s in the Delta, where it was regarded as being cheap and fairly effective (Wilkins and al-Sawy 1977). As we shall see in the next chapter, the program was strengthened in 1988 with the introduction of a new, more effective, and safer drug, praziquantel.

What happened to women and girls in the Egyptian record?

Most early descriptions of schistosomiasis in Egypt were by men, and about men. They illustrated the way a male gendered dialogue about schistosomiasis developed. Records from the Qasr al-'Aini Hospital in Cairo for 1900, for example, show that only ten per cent (6 of 58) of the "medical and surgical cases" involving schistosomiasis were women (Keatinge 1927: 200–1). Similarly, for a number of years before 1907 94% of the 1,346 cases of schistosomiasis admitted to that hospital were males (Madden 1907: 16). Although the actual number of cases among males was likely to have been greater than among females, such gender contrasts were exacerbated by the gendered difference in treatment-seeking behavior, with males more likely to seek and obtain treatment than females.

Frank Madden, the expatriate dean of the Medical School and hospital during this period, was well aware that schistosomiasis was a serious problem. Commenting on the years between 1907 and 1909, he found that 10% of the people admitted for treatment were suffering "from the last stages of the pathological destruction of the kidneys...bladder, urethra and rectum produced by severe and repeated bilharzial infection" (Madden 1919:293).

In 1898, Madden published the first description of what is now known as female genital schistosomiasis, complications arising from the deposition of eggs in the genital tract (Madden 1898). The account was based on only a very few cases. Indeed, it appears that Madden and his colleague, Frank Milton, saw only three or four victims of the disease at Qasr al-'Aini Hospital and all were suffering from advanced genital disease. Neither he nor his contemporaries knew anything about the earlier stages of the disease. However, Madden's description is important in the light of the recent identification of significant reproductive morbidity due to female genital schistosomiasis in a rural community in Egypt (Talaat 2001).

Madden noted, in 1907, that "for some reason not yet understood," cases of schistosomiasis were rare among women and girls. He suggested that, *if* the disease causal agent (still unknown at the time of writing) entered the human body through the skin, the difference between infection rates in men and in women could be explained by differences in exposure to infected water during farming and bathing.

Once Leiper had identified the schistosomiasis transmission cycle, and control strategies began in the 1920s, women and girls again seemed to disappear from the record. In most accounts farmers (assumed to be male) were identified as the most affected sector of the population. The role of their wives and daughters, who frequently worked alongside them in the fields, or as day-laborers, was overlooked. The frequently reproduced poster, prepared for the schistosomiasis control program in 1924, shows a wife lamenting over the prone figure of her schistosomiasis-stricken farmer husband, stretched out beside an irrigation canal (Farley 1991, dustcover; Jordan 2001).

By the 1930s, epidemiological studies of schistosomiasis enabled researchers to calculate infection rates among women, men, and children, in the settlements in which they lived, rather than relying, as earlier, on information about hospital patients who were in the last stages of the disease. Dr. J. Allen Scott, who in 1935 conducted the first national survey of schistosomiasis, attempted to obtain a representative sample of males and females in the national sample for his house to house survey. He noted with some satisfaction that 50.7% were female, close to the sex ratio found in the national census figures of the time. However, when it came to analyzing his data, Scott (in line with current thinking in his day) focused on the contrast in infection rates between areas of basin and perennial irrigation. He only briefly mentioned the lower overall rates for females (Scott 1937: 600–5). In short, having surfaced briefly in his work, women were again submerged.

A late 1970s study of morbidity in the southern Delta governorate of Qalyubia focused on adolescent boys, who were known as a group to have high infection levels and egg counts (the measure of the "intensity" of infection). The study identified young males with high egg counts, who were made to spend a brief period in hospital so that a battery of tests could be administered to identify pathology due to schistosomiasis. The study found considerably more serious illness associated with *S. haematobium* (mainly pyuria, proteinuria and hematuria), than with *S. mansoni* (Pope et al. 1980). No comparable study of female morbidity has been attempted in Egypt, perhaps because of the greater problems likely to be encountered in persuading parents to allow their young daughters to go to hospital for tests.

Few researchers provided evidence, in their published reports, that they were aware that the way in which they carried out their studies could affect their findings with respect to gender issues. Helmut Kloos (1983; 1990) recognized that local cultural constraints would make the study of the water contact behavior of girls in Upper Egypt especially difficult for a predominantly male research team. Yet, with this and one or two other exceptions, most epidemiologists in Egypt overlooked the significance of differences in male and female infection rates, even in the 1990s.

5

Schistosomiasis in Egypt: 1988–96

Praziquantel: a new tool for schistosomiasis control in Egypt

Egypt, with its long history of research and schistosomiasis control activities, was in 1988 the first country in the world to adopt the new drug praziquantel as the mainstay of its national treatment and control program. At the same time, the Ministry of Health launched a collaborative Schistosomiasis Research Project (SRP) to provide information to support the program. Projects supported by the SRP led to an explosion of (mainly epidemiological) information about the status of the disease in the early 1990s. In this chapter we present those research findings which provide a national context for our own local level research work. Our study, initiated in 1991, also funded by the Schistosomiasis Research Project, used a rather different approach from other SRP projects. Before describing it, however, we need to explore some of the events leading up to the introduction of praziquantel in 1988.

From the late 1960s onward, as the number of people afflicted with schistosomiasis continued to grow, public health officials in Egypt realized that the long courses of injections required by tartar emetic and its derivatives could not solve the problem. They turned to new drugs, such as oxamniquine and metrifonate, and, later, praziquantel, which promised higher cure rates and fewer side effects. WHO, UNICEF, the World Bank, and pharmaceutical companies supported an international program to assess the effectiveness of these various drugs.

By 1980, based on studies in Egypt, South Africa, and Kenya, researchers had identified the single dose drug praziquantel as the most suitable antischistsomal drug. They found that it was safe and effective, and could be used for both *S. mansoni* and *S. haematobium*. As a result of these studies, in 1984 the WHO Expert Committee on the Control of Schistosomiasis endorsed the use of praziquantel to control morbidity due to schistosomiasis, as part of an integrated strategy of treatment and prevention (WHO 1993: 1).

The cure rate for praziquantel is now estimated to be over 80% for *S. haematobium*, and between 60% and 80% for *S. mansoni*. Treatment for *S. haematobium* has been found to reduce proteinuria (protein in the urine), hematuria, urinary iron loss, and leukocyturia (abnormal levels of leukocytes, cells, in the blood). Half of all *S. mansoni* patients who were treated experienced a reduction in liver and spleen size (El Khoby et al. 2001: 12, 14). These results were achieved after patients had taken a single dose of praziquantel orally. The recommended dose was 40 mg. for every kilogram of body weight, which for an adult meant three or four 600 mg. tablets. Praziquantel was particularly valuable in Egypt because it could be used against both *S. mansoni* and *S. haematobium*.

Children, in Egypt as elsewhere, had been shown to have high infection rates and high worm burdens. The Ministry of Health, which already provided screening and treatment for school children, set about identifying the most appropriate way to incorporate the new drug into their program. In 1983 a pilot program was established in two highly endemic districts of Beheira governorate, south of Alexandria. The purpose of the project was to assess the impact of a single treatment on a population of school-age children. An initial prevalence survey assessed the situation before praziquantel was administered; the second survey followed one year later, and the third assessed the situation the next year (no praziquantel having been administered between these two surveys). *S. mansoni* infection rates fell from 60.3% before treatment to 24.8% one year later, increasing to 41% the following year. The corresponding figures for *S. haematobium* were 37.6% before treatment, 5.5% one year later, and 9.9% two years later (Spencer et al. 1990). These findings reflected both the greater effectiveness of praziquantel against *S. haematobium,* and the fact that many children were being reinfected. Although the decline in infection rates immediately after the administration of praziquantel was dramatic, it was clear that repeated treatments would be needed to make a real impact on overall infection rates.

The Beheira governorate researchers estimated that they had only been able to screen about 70% of all school-age boys, and an even smaller proportion, about 30%, of school-age girls (El Malatawy et al. 1992). But they apparently did not see this discrepancy in coverage rates between boys and girls as a potential problem for a school-based program. As we will see in chapter 11, for a number of reasons programs in operation in the early 1990s did not reach girls as effectively as boys.

The National Schistosomiasis Control Program: objectives

During the thirty year period when laboratory scientists were working on a new generation of anti-schistosomal drugs, and epidemiologists were field testing them, the government of Egypt had not been standing still. In 1977 (eleven years before the official adoption of praziquantel) the National Schistosomiasis Control Program (NSCP) was launched. Its main objective was to "reduce schistosomiasis to a level at which it is no longer a public health problem" (El Khoby et al. 2001: 9–10). The Program eventually covered all the rural areas of the Nile delta, the valley south of Cairo, Fayoum, and the newly settled agricultural areas.

In 1988, at the same time that the Ministry of Health officially incorporated praziquantel into the National Control Program, it modified other aspects of its overall program. Among other things, it modernized its approach to the control of snail vectors. It also gave new emphasis to preventing the spread of schistosomiasis in newly settled areas (such as the eastern and western fringes of the Delta and around the Suez Canal). Also emphasized after 1988 was the prevention of the spread of *S. mansoni* in Upper Egypt, where an increasing number of cases were being found. The revitalized Program also strengthened its earlier strategy to target vulnerable groups, especially school children. The scope of the Program, as of 1988, is indicated in table 5.1.

Table 5.1: The scope of the Egyptian National Schistosomiasis Control Program, 1988

1. Free diagnosis and treatment for:
   all school children, with periodic screening at rural health facilities;
   all adults attending rural health facilities
   all adults who, in village sample surveys of around 10% of residents a month, were found positive;
2. Snail control through the application of niclosamide to canals in villages with over 20% prevalence, and in canals where infected snails were found;
3. Health education through the Health Education Department of the MOH and through the mass media;
4. Liaising with other authorities to improve water, drainage and irrigation conditions.

The main activity of the Program after 1988, as had been the case since 1977, was free diagnosis and treatment administered through the national primary health care system. These services were delivered chiefly at the 2,400 rural health units in the Nile valley and the Delta. The Program targeted Egypt's 15 million school children between the ages of 6 and 14, the years of compulsory, free schooling. In addition, all adults attending the health centers were required to undergo testing for both types of schistosomiasis, regardless of the symptoms they presented. The MOH estimated that in 1995, 3,500 technicians at MOH units tested 22 million stool and urine samples (El Khoby 1995). In chapter 11, we will discuss the stool testing facilities we found in operation in the early 1990s in our study communities.

After 1988, as in the years after 1977, canals continued to be treated to destroy snail vectors. However, two important changes were introduced. Blanket mollusciciding was replaced by focal treatment restricted to canals in highly endemic villages, and at sites where infected snails were found. Copper sulfate was gradually phased out and replaced by niclosamide, an agent that was seen as less toxic to the aquatic environment. Focal treatment was also advantageous as it required the use of far smaller quantities of molluscide, at a time when the cost of niclosamide, a petroleum derived chemical, was rising sharply.

In 1988 the health education program was expanded, and focused on the mass media, mostly in the form of TV messages. The MOH pledged to continue to work with other government agencies to improve water, drainage, and irrigation conditions (see El Khoby et al. 2001).

The organization of the Program

In this section we will describe the National Schistosomiasis Control Program as found in operation during our research in Munufiya governorate, beginning in 1991. As the Program undertook a number of activities designed to control the disease, both curative and preventive, it could claim to be an integrated program, following the recommendations of WHO.

Overall responsibility for the NSCP rested with the Ministry of Health in Cairo. The Ministry was a highly centralized organization, although, at the time of our study, it was developing a strategy for decentralization. For the sake of clarity it is best to regard the Program as a conglomerate that contained more than one chain of command; depending on circumstances, one chain sometimes overlapped with others. For some activities the chain of command was clearly vertical (top-down) from the central Ministry of Health to regional (governorate), to the district (*markaz*) and to the local village level. But for other activities at the intermediate level, the top-down linkage was supplemented by a horizontal linkage. In these cases, the Schistosomiasis Control Program in the governorate was obliged to cooperate with other Units at the same level, such as health education, in order to provide certain schistosomiasis services, as shown in figure 5.1.

Figure 5.1: The organization of the National Schistosomiasis Control Program

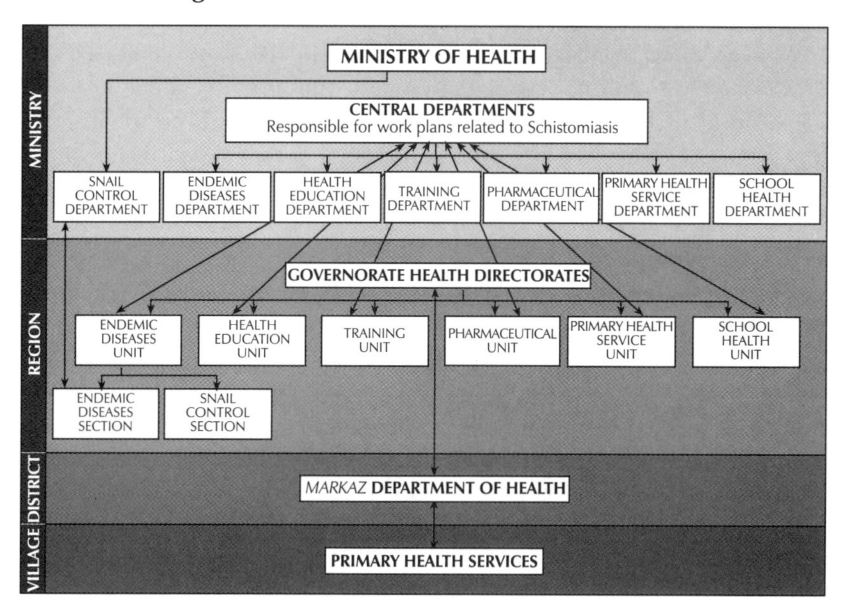

Within the Ministry of Health in Cairo, the Endemic Diseases Department had specific responsibility for the NSCP. Other Departments in the center that collaborated with the work of the NSCP included, for example, the Health Education Department and the Training Department. Each Department in the Ministry prepared the work plans for the corresponding like-named Units at the governorate level.

At the governorate level (as in Munufiya governorate), the Schistosomiasis Control Program was situated in the Endemic Diseases Unit. When the governorate Program required support activities such as health education or training, staff had to request the support from their sister Units in the governorate, such as the Health Education Unit or the Training Unit.

The Schistosomiasis Control Program at governorate level was responsible for collecting all reports from health personnel at the *markaz* (district) level and sending them to the NSCP office in the Endemic Diseases Department in Cairo. An exception to that rule was the Snail Control Section, which sent all findings directly to the Ministry of Health in Cairo.

Still at the governorate level, a coordinating body—the High Committee for Schistosomiasis—had a mandate to meet monthly under the chairmanship of the Governor, and to report to the governorate Health Directorate. Members of the schistosomiasis committee included representatives from the Ministries of Education, Public Works and Water Resources, Housing, and Information, and religious leaders.

Moving down the hierarchy, all the Units in the Governorate Health Directorate communicated directly with the *markaz* (district level) Department of Health; this was a single department, with no subsections. The *markaz* Department of Health sent reports to the relevant Units in the Governorate, such as the SCP in the Endemic Diseases Unit. The *markaz* Department of Health was responsible for providing primary health services at the local level.

At the local level, MOH facilities could link up with other departments that had an interest in schistosomiasis control through the Executive Village Council and elected representatives from various villages and subvillages served by the Council. Diagnosis and treatment was provided free for the local population, as well as for school children.

In 1992, overall responsibility for the school screening program was given to another governmental organization, the Health Insurance Organization (HIO). It was intended that the HIO should operate independently, without collaborating with the rural health units. However, as of 1996, the HIO was not yet operational in the study area. Other changes in the organization of the National Program have also been introduced, such as mass treatment (without prior diagnostic testing) in highly endemic areas. However, this too had not yet come into effect in the Munufiya study villages during the time of our research there. The impact of these changes will be considered in a final chapter, as part of an assessment of the relevance of our study to the situation as it exists in Egypt at the beginning of the new millennium.

The Schistosomiasis Research Project

In 1988, in the same year that the Ministry of Health began to use praziquantel in its national control program, it launched the Schistosomiasis Research Project (SRP). This multi-million dollar project was undertaken in collaboration with USAID and Egyptian universities. The stated objective was to provide information and guidelines for the newly revised praziquantel-based control Program. A major component of the Schistosomiasis Research Project was a survey of the disease in nine Delta and Nile valley governorates. In addition, the SRP also asked research teams at Egyptian universities to submit proposals on topics relevant to the MOH control program.

The nine-governorate survey was called EPI 1, 2, 3, because of its three major goals. The first was to provide an overview of the distribution and intensity of infection. The second was to identify and attempt to explain the differences in levels of schistosomiasis infection between villages. The third goal was to measure the severity of the clinical pathology due to the disease (morbidity).

In an attempt to get a balanced picture of infection levels in each governorate, EPI 1, 2, 3 used a specially designed multi-stage sampling frame. On the basis of this sample, researchers felt confident to make generalizations about the whole population of these governorates.

The initial survey, conducted in each governorate in 1992, provided information on infection rates at that time ("prevalence" in epidemiological terminology) and intensity of infection (measured by the number of schistosome eggs in urine or stool specimens) for both *S. haematobium* and *S. mansoni*. Subsequent annual surveys identified infections that had occurred since the previous round of tests ("incidence").

The 1992 survey initially covered 17,172 households in 251 rural communities. It surveyed slightly more females than males, 45,874 compared to 45,176, a total of more than 90,000 people. In the light of this achievement, the survey was said to be the largest community-based epidemiological study to have been conducted anywhere in the world (El Khoby et al. 2000 A; Hussein et al. 2000).

Main findings of the 1992 survey: distribution and intensity of disease

The 1992 survey provided baseline data on the prevalence and intensity of infection for both types of schistosomiasis in nine governorates, five in the Delta and four in the Nile valley south of Cairo (see page xxii for a map of governorates). It confirmed that *S. haematobium* was the main form of the disease in Upper Egypt, and that *S. mansoni* now predominated in the Delta. These findings, showing the predominant form of schistosomiasis at the governorate level, are summarized in table 5.2.

In addition to indicating the regional distribution of each type of schistosomiasis, the findings also showed that the prevalence of *S. haematobium* in Upper Egypt was lower than that of *S. mansoni* in the Delta. The prevalence rates of *S. mansoni* in the Delta were actually higher than those found by earlier surveys and by MOH records.

Table 5.2: Summary findings for nine governorates, 1992

	Total	Male	Female	Odds ratio	Community prevalence
Lower Egypt (the Nile Delta): *S. mansoni*					
Kafr al-Sheikh	39.3%	55.3%	36%	1.44	24.5—68.9%
Gharbia	37.7%	46.9%	37.9%	1.45	17.9—79.5%
Munufiya	28.5%	34%	21.1%	1.93	0.3—72.9%
Ismailia	42.9%	50.5%	36.1%	1.81	12.3—71.3%
Qalyubia	17.5%	25.3%	16.9%	1.67	2—45%
Upper Egypt: *S. haematobium*					
Fayoum	13.7%	16%	11.5%	1.46	0—27%
Minya	8.9%	12.1%	5.9%	2.19	1.9—32.7%
Asyut	5.2%	8.2%	4.3%	1.96	1.5—20.9%
Qena	4.8%	7.1%	3.8%	1.94	0—20.6%

Source: Summarized from papers in the *American Journal of Tropical Medicine and Hygiene*, 2000, 62 (2) Supplement.

However, a few people with *S. mansoni* were found in all the Upper Egyptian governorates surveyed. The overall level of *S. mansoni* was higher in Fayoum governorate, just south of Cairo, than in the other Upper Egyptian governorates. In Fayoum, the fact that three of the four hamlets recording high *S. mansoni* rates (between 25 and 35%) also recorded low *S. haematobium* infection rates (<7%) suggested that *S. mansoni* might actually be replacing *S. haematobium* as the major type of schistosomiasis in those communities (Abdel-Wahab et al. 2000 A and 1993).

In the other three Upper Egyptian governorates surveyed, the *S. mansoni* cases were concentrated in a small number of villages. This finding was alarming as it suggested that active transmission of *S. mansoni* was now taking place there. In earlier surveys, *S. mansoni* had only been found among migrants or visitors from *S. mansoni* endemic areas to the north (El Khoby et al. 2000 B; King et al. 1982).

Overall, the 1992 survey found a significantly higher rate of infection among males than among females. For *S. haematobium* in Upper Egypt, the infection level was 5.6% for females and 10.1% for males, with an odds ratio of 1.91 (the odds ratio measures the strength of the correlation between two variables, in this case infection and a demographic variable; the higher the odds ratio the stronger the correlation). In the Delta, gender differences in infection with *S. mansoni* were somewhat less marked, 31.4% for females and 41.5% for males, an odds ratio of 1.56.

The 1992 survey confirmed earlier findings that young people under the age of twenty had the highest infection rates. For example, in some villages in Fayoum the odds for infection with *S. haematobium* among children and young people were double or triple those for adults. Here, almost 80% of those infected with *S. mansoni* were aged 20 or younger (Abdel-Wahab et al. 2000 A).

Overall, the survey recorded that the peak infection rates occurred at a younger age for *S. haematobium* than for *S. mansoni*. In Kafr al-Sheikh governorate, in the northern Delta, fairly high rates of infection for *S. mansoni* continued into middle age (Barakat et al. 2000). In Ismailia, on either side of the Suez Canal, *S. mansoni* infection levels peaked among those aged 20 to 30, but were higher than elsewhere until age 55; this pattern was thought to be the result of the relatively recent introduction of the disease into this newly settled area (Nooman et al. 2000).

In all governorates, the differences in prevalence rates from village to village confirmed that the infection was highly localized. For example, in the delta governorate of Munufiya, with an overall prevalence of 28.5%, the prevalence of *S. mansoni* in 27 rural settlements ranged from 0.03% to 79.5%. However, no SRP researcher made a serious attempt to identify local conditions that might have accounted for village-to-village contrasts, beyond noting that hamlets usually had higher infection rates than the larger villages.

Technicians in SRP projects used the modified Kato-Katz thick smear technique for fecal specimens and Nucleopore filtration for urine to count eggs in fecal or urine specimens, rather than simply noting the presence or absence of eggs (Cline et al. 2000). Thus, the 1992 survey provided information on the intensity of infection; intensity is an indicator based on the number of eggs expelled by an infected individual per gram of stool or 10 ml. of urine. A composite measure of intensity for the whole infected population or a subgroup of that population is provided by the Geometric Mean Egg Count (GMEC). Intensity provides epidemiologists with an approximate indication of morbidity, and of the potential for disease transmission.

The EPI 1, 2, 3 survey found that, for both forms of schistosomiasis, and in all age groups, intensity was higher for males than females. For children between 5 and 14 years of age with *S. haematobium*, the intensity of infection was higher than for any other age group, with a GMEC of just over 10 eggs per 10 ml. of urine. Adults over 25 in different age and gender groupings recorded GMECs around half those recorded for children. For *S. mansoni*, the intensity of infection was highest among young people, specifically for children between the ages of 10 and 14 (El Khoby et al. 2000 B).

In some governorates, such as Kafr al-Sheikh and Munufiya, the survey researchers considered that the overall low level of intensity recorded in 1992 was the result of effective diagnosis and treatment at rural health units (Abdel-Wahab et al. 2000 B; Barakat et al. 2000; El Sayed et al. 1997). It is known that most patients who become reinfected after treatment have lower egg counts than in previous infections.

The low overall egg counts revealed in the EPI 1, 2, 3 surveys may have resulted in an underestimate of the overall infection rates for the nine governorates. This underestimate occurred because all surveys relied on a single fecal or urine sample for diagnosis. As egg output is variable over time, both within a 24 hour period and from day to day, some low intensity cases (with a smaller number of eggs excreted) were likely to have been missed. Thus, in Kafr al-Sheikh governorate, the principal investigator estimated that the actual prevalence in the 1992 survey was probably at least 50%, although the overall corrected prevalence in the sample survey (based on a single Kato-Katz test) was 39.3% (Barakat et al. 2000). However, as all parasitological testing for EPI 1, 2, 3 and all other SRP sponsored projects was standardized, the overall comparison of rates in different areas of Egypt would not have been affected.

Identifying those at risk of infection

Following standard epidemiological practice, in the 1992 survey the infection rates of different categories of people were calculated in order to identify "risk factors" for infection. These "risk factors" were based on a statistically significant correlation between a certain "factor" and infection. The "risk factors" identified in the 1992 survey were similar for both kinds of schistosomiasis. Overall, males had a significantly higher infection rate than females, as did young people aged 5 to 20 (except in Ismailia governorate). Thus, being male was identified as a "risk factor," as was being under the age of twenty.

The EPI 1, 2, 3 interviewers asked individuals if they had come into contact with canal water. Three different groups who reported specific kinds of canal contacts were identified as "at risk": males who had entered the canal to bathe, children under 15 years of age who reported that they had played in the canal, and women who had reported "washing" in the canal—the kind of washing was not specified. The survey also categorized as "at risk" people who had reported that they had been previously infected or treated for schistosomiasis, or who had had symptoms of the disease at some time.

The survey also identified people living in hamlets (*'izab*) as "at risk" because they had significantly higher rates of infection than those living in villages. The researchers suggested that, compared to inhabitants of larger settlements, these people lacked safe water and sanitation and were more likely to come into contact with canal water; they also lacked easy access to facilities for diagnosis and treatment. However, they did not follow up these suggestions.

EPI 1, 2, 3 researchers also were unable to identify any link between an individual's educational status and infection. However, they made no comment about the possible reason for their finding. They probably had originally hypothesized that people who had little or no education would be more likely to use the canals on the grounds that they may not know of the disease danger, or that they were poor and more likely to live in houses without safe water or sanitation.

The interviewers asked individuals over fifteen years old about their main occupation. This data on the primary occupation, when correlated with infection status, apparently did not indicate any of these occupations as a risk factor (El Khoby 2000 B). Thus, in spite of the emphasis in earlier epidemiological studies on farming and irrigation as associated with infection, the EPI 1, 2, 3 survey could not specifically identify farmers as being at risk. It is likely that they failed to do so because they did not attempt to identify the many part-time farmers, women and children as well as men, who worked in the fields after their regular working day and on Fridays. In chapter 9, we will explore this issue as it relates to our two Munufiya study villages.

Morbidity due to schistosomiasis

Doctors practicing in Egypt often assumed a connection between certain of their patients' conditions and chronic schistosomiasis infection. For example, by the 1970s physicians in the Nile delta were reporting long-term morbidity such as high blood pressure, portal hypertension, and death from hematemesis (internal bleeding), which they considered were related to the increasing rates of *S. mansoni* in the Delta (El Khoby et al. 1998). However, it is difficult to assess the precise extent and severity of these conditions and their relationship to schistosomiasis infection from such reports. Large scale studies that could identify various kinds of morbidity and relate these to current or previous infection with schistosomiasis were clearly needed.

Aware of this problem, the EPI 1, 2, 3 1992 survey pioneered the use of a new technique to identify morbidity in the community, using a new type of ultrasound machine that was portable. Brought into the village, this machine could identify morbidity among local people at an early stage of the disease.

The heavy ultrasound equipment used previously could not be easily moved about. Suitable only for use in a hospital, its main function was to identify more serious morbidity and disease conditions that were already well advanced.

The results of these EPI 1, 2, 3 community-based ultrasound examinations were compared to results for the same individual obtained from physicians' clinical examinations and the individual's disease history (Abdel-Wahab et al. 2000 C). The findings in some cases seemed to establish a clear cause and effect relationship between the presence of schistosomiasis and certain conditions, while in other cases the results were more ambivalent.

In Munufiya governorate, clinical and ultrasound examinations identified a clear association between peri-portal fibrosis and infection with *S. mansoni* in adults (but not in children). Here, ultrasound was found to be more effective than routine clinical examinations in identifying enlargement of the liver and spleen due to *S. mansoni* (Abdel-Wahab et al. 2000 C; El Khoby et al. 2000 B). Later, a large-scale study in Fayoum governorate compared ultrasound data collected in 1997 to that of the 1992 survey collected five years earlier. The 1997 data indicated a significant decrease in the prevalence of both hepatosplenic (liver and spleen) and urinary tract morbidity over the five year period. El Khoby (2001) suggested that these findings were related to the decrease in cases of both types of schistosomiasis in the governorate during that time.

Many researchers have noted the relationship between bladder cancer and chronic, heavy, and repeated infection with *S.haematobium*. Detailed information about the different kinds of cancer in Egypt is now available from cancer registries in the two largest cities, Cairo and Alexandria, and from the records of cases treated in the National Cancer Institute in Cairo. These data show that the pattern of cancers in Egypt is very different from that in most high income countries, where lung, colonorectal, and breast cancers are the three most frequent malignancies. In Egypt, as of the early 1990s, bladder cancer was the most frequently recorded type of cancer, responsible for around one quarter of all malignant tumors, thus reinforcing earlier conclusions about the relationship between bladder cancer and *S. haematobium* infection.

Similarly, statistics collected *after* the introduction of universally available praziquantel treatment in 1988, suggest that the number of new cases of squamous cell carcinoma, the type of bladder cancer associated with *S. haematobium*, was in decline, compared with the number of new cases reported earlier in the decade (El Khoby 2001). Further confirmation of these findings might be possible if the cancer records could be analyzed according to the patients' rural or urban residence, as almost all schistosomiasis is transmitted in rural areas.

Another legacy of the pre-praziquantel era (before 1988) has been pinpointed in a recent study that shows that standard injections with tartar emetic were likely to have been responsible for the current high rate of infection with the blood borne hepatitis C virus (HCV). In Egypt the level of HCV is 10–15%, one of the highest in the world. At the time when tartar emetic injections were being used for schistosomiasis, public health specialists were not aware that the standard sterilization procedures for the injection equipment could not protect patients from viral infections. These could be transmitted during mass treatment that required a series of tartar emetic injections over several weeks. The authors of the study suggested that the greater number of men and adolescent boys receiving injections for schistosomiasis could explain why the majority of cases of HCV in Egypt in the 1990s were men over 40, who had received injections 20 or more years ago (Frank et al. 2000).

What we need to know about gender differences in infection

Beginning in the 1930s, epidemiological reports of infection in rural communities found more males than females infected with schistosomiasis. In the 1960s, in villages in Beheira governorate, just south of Alexandria, Farooq and his colleagues found that infection rates for both types of schistosomiasis were higher among males than among females for all age groups except for those under five years of age. At a time when farming was still a full-time occupation (rather than a part-time activity as it has become now), Farooq identified farmers and farm laborers as the occupational group with the highest infection rates for both *S. mansoni* and *S. haematobium*; 43% for the 2,842 female farmers and 52.8% for the 2,900 males. Farooq's survey indicated an approximately equal proportion of men and women were farmers, just under half of the women and half of the men surveyed (Farooq et al. 1966A).

In the 1970s and 1980s, a number of researchers noted a bigger gap between the infection rate of females and males in Upper Egyptian communities than in the Delta (Mansour et al. 1981; Miller et al. 1981; Hammam et al. 1975; King et al. 1982). In 1982 C.L. King and his colleagues suggested a reason for this gender gap. They wrote that: "It is frequently said that the females of Upper Egypt have had less exposure to contaminated water than their northern counterparts due to different working patterns and more conservative behavior." (King et al. 1982: 326)

The 1992 nine-governorate survey found that more males than females were infected in all governorates. They also found that, overall, the difference *between* male and female infection rates in the four Upper Egypt governorates (an odds ratio of 1.91) was greater than in the five Lower Egyptian governorates (an odds ratio of 1.56). On the governorate level, differences between male and female prevalence rates were highest in Minya (12.15% for males and 5.9% for females, an odds ratio of 2.19), a governorate endemic for *S. haematobium*. They were lowest in the Delta governorate of Kafr al-Sheikh (55% for males and 36% for females, an odds ratio of 1.44), where overall prevalence rates were highest and only *S. mansoni* was found.

In Minya, the association between infection and the three kinds of canal exposure identified by EPI 1, 2, 3 were relatively low: an odds ratio of 1.67 for males bathing, 2.10 for children playing and 1.48 for females washing. The comparable figures for Kafr al-Sheikh were 3.45, 3.74 and 3.09 (Barakat et al. 2000; Gabr et al. 2000). This suggests that in Minya (with an overall prevalence rate of 8.9% compared to 39.3% in Kafr al-Sheikh) many water contact activities took place at sites where there were few or no vector snails or schistosomes. In contrast, in Kafr al-Sheikh, where vector snails and schistosomes were much more widely distributed, the correlation between water contact activities and infection was much higher for all three "at risk" groups.

These kinds of comparisons suggest the need to look more closely at the relative risk of different kinds of exposure activities for women, men and children, in various parts of Egypt. While the 1992 survey provided information on the number of males and females infected, and on certain behaviors that could be interpreted as "risk factors," it did not explore these questions in greater depth. There is clearly also a need to find out how people *actually* behave, in contrast to the way *they say* they behave when confronted by an interviewer.

Overall, the various constituent parts of the Schistosomiasis Research Project, including the EPI 1, 2, 3 study, prompted many questions about gender differences in infection (and indeed about other aspects of gender and schistosomiasis). The absence of any discussion about gender in EPI 1, 2, 3 and most other SRP studies is probably the result of the near absence of anthropologists or other social scientists working in health related topics in collaboration with biomedical and public health colleagues. When SRP made a special request for proposals on socio-economic topics, few local social scientists responded.

Only two of the projects accepted by SRP had social scientists as principal investigators—both were carried out under the auspices of the Social Research Center at the American University in Cairo. Only ten of the 103 completed SRP projects were on socioeconomic topics. The only studies that used the rich database provided by EPI 1, 2, 3 focused on the coverage of screening for school-age children. One study found that children (especially girls) who did not attend school were more frequently infected than boys and girls who went to school (Husein et al. 1996). Concerned by these findings, the researchers then designed and tested a strategy to reach children who were not in school (Talaat et al. 1999A). This study, which shares our concern for gender equity, will be discussed in chapter 11.

The epidemiological studies presented in this chapter, particularly EPI 1, 2, 3, give a broad picture of the distribution of the two forms of schistosomiasis in Egypt in the early 1990s, of the characteristics (such as age and sex) of those infected, and of "risk factors" for infection. As such, these studies reflect a biomedical and epidemiological view of disease and disease agents. They were not designed to study gender and human behavior in detail, or their relationship to schistosomiasis infection. They do, however, pose questions about gender and real-world risk behaviors. To these topics we now turn.

6

Research Concepts, Methods, and Procedures

A participatory, interdisciplinary study

This chapter will briefly describe the research methods and procedures we used to carry out a gender-conscious study of schistosomiasis, principally in two Nile delta villages. In any setting other than late twentieth-century Egypt, these villages would be termed "small towns" or "semi-rural" communities due to their size—each had more than 7,000 inhabitants—and the fact that the majority of households no longer regard farming as their main means of support. However, as local people continue to call them "villages" (*qura),* and think of themselves as "rural" people in comparison to "townspeople," this is the designation we too will use.

In Chapter One we introduced the research project and characterized it as being participatory, action-oriented, and community-based. This approach provides scope for a consciousness of gender that is often absent from more structured studies. In the present chapter we will begin by enlarging on these themes and then incorporate them into a model for a holistic village-level study. More detailed information about the methodology of the study will be found in the Appendix.

The two co-authors—a cultural anthropologist and a geographer—worked with two sociologists and an anthropologist (also at the Social Research Center of the American University in Cairo) who helped with the community research. Our team also included experts in biomedicine, epidemiology, and environmental health based at the High Institute of Public Health in Alexandria, who organized the epidemiology studies and the collection of information about water quality in canals and domestic water sources, and about snail populations in the village canals. Thus, as an interdisciplinary team, we shared concerns about gender and behavior while at the same time drawing on findings from the biomedical sphere, epidemiology, and environmental health.

The participatory approach we used is familiar to anthropologists, who are accustomed to working with local people in order to come to a sympathetic understanding of their view of their life and activities. Oliver Razum and colleagues (1997) differentiated action research from operational research, which usually involves asking a restricted range of questions about a particular health program, and in which local people are only involved in data gathering. For us, "action research" was an on-going exploration that entailed, as far as possible, the continuous involvement of local people (El Katsha et al. 1993–94; see also Chambers 1997; Nichter 1994; Smith et al. 1993).

Our starting point was to look at what was happening in families and households in the study communities, working closely with local people themselves. Unraveling and exploring the dynamics of schistosomiasis transmission can best be done at the local level, for it is here that people learn to live with the disease, or to avoid it. It is here that family members decide that they have become sick, and that they should seek diagnosis and treatment.

A model for a holistic village level study

To illustrate the scope and orientation of our research, we developed the model shown in figure 6.1. The model presents patterns of human behavior in a spatial context. It starts with the understanding that village residents experience life as individuals in a household, with women, men, and children sharing more or less the same living space, and interacting with each other on a daily basis. At the same time, interacting with the larger community, they use available village-level services and learn to cope with the shortage of potable water, polluted canal water, and unsafe sanitation. At the local level, government employees responsible for providing and maintaining these services act within the constraints of time, finance, training, equipment, and their own perceptions of the village setting.

Our model extends beyond the village to include regional and national government authority. It was (and is) at these levels that the decisions are made (or not made) to provide communities with basic services which are necessary if schistosomiasis is to be brought under control. The regional and national authorities set policies and limits, within which local staff must operate. We found that each of the government departments involved directly or indirectly in schistosomiasis control, and their local employees, had a their own set of responsibilities and their own ways of approaching and executing those responsibilities.

Figure 6.1: A model for a holistic village-level study

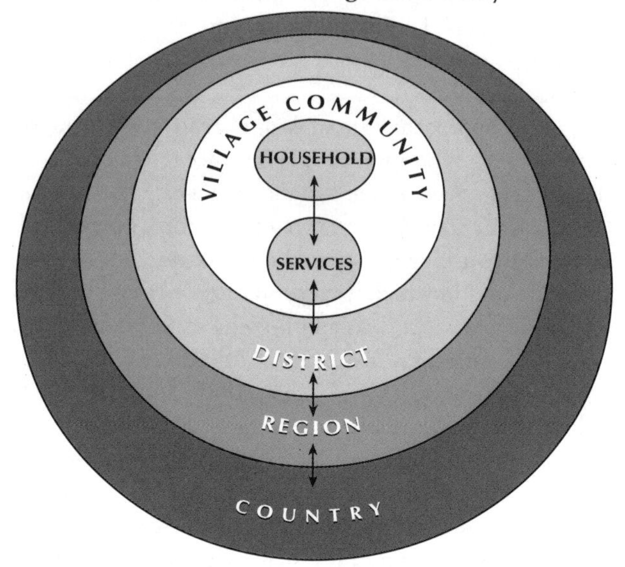

Source: El Katsha and Watts 1997.

Our model is unusual in that it includes many local services (and local employees) that are responsible for disease control; it also looks at the way they interact with district and higher level authorities. Staff involved in these interactions are responsible to many different ministries—including those concerned with health, education, water supply, irrigation, and local government. Thus, our model includes processes that operate on various levels; it focuses on what *actually* happens in practice, in contrast to a unified model which presents a top-down view of what planners consider *ought* to happen (El Katsha and Watts 1997; van Ufford 1993).

The household as a unit of study

The household lies at the center of our model, for it is here that the individual (in sickness and in health) lives out his or her daily existence. Depending on individual circumstances and phases in the life-cycle, it is within a household that most people (as infants progressing into childhood, young adulthood, maturity, and old age) have their being and within which they develop their own sense of identity. The household is usually defined as a unit of consumption and in poor farming societies as a unit of production; members eat from a common pot, although they may not eat at the same time or in the same place.

The household can also be defined in terms of the bounded, physical space occupied by its members— often identified as a "private" space as opposed to a "public" space outside. Health researchers have found it useful to identify these as separate "domains" of disease transmission, associated with different patterns and scales of social interaction. Some diseases have been characterized as being transmitted outside the household, while others are directly transmitted among household members and clustered within certain households (Cairncross et al. 1996).

In our study area, the term *'ila* is the colloquial Arabic term that encompasses the concept "household" but it can also refer to the extended kin group, beyond the single residential unit. In practice, however, when used about interactions within a community setting, it clearly refers to the smaller actual residential unit. In English usage, the word "home" has a similar ambiguity, but its meaning can clearly be identified by the context within which it is used. The classical Arabic term *usra*, which is used in the national census and in many epidemiological studies, can also refer to a "household."

In our study area, as in most societies, it is within the household that children are socialized. As we will see, with regard to concepts about water and about sanitary matters, it is within the household that infants and young children (boys and girls) first acquire their understanding about which behaviors are appropriate and which are not. Thus, the household can also be seen as a setting for reproducing gender roles and transmitting them to the next generation (World Bank 2001: 151–54).

In the 1990s, social scientists and public health specialists began to explore ways in which the household could provide an integrating framework for the study of health, especially in low income countries. The "household production of health" has been defined as "a dynamic behavioral process through which households combine their (internal) knowledge, resources, and behavioral norms and patterns with available (external) technologies, services, information, and skills to restore, maintain, and promote the health of their members" (Berman et al. 1994: 206). The emphasis on the "dynamic behavioral process" involves looking at health and health outcomes within the context of gendered interaction between spouses, between siblings, between senior women and their daughters-in-law, and between fathers and sons. At the same time, the model incorporates external factors, such as health facilities and other public services that affect the health and well-being of local residents.

Identifying the study communities

When we began our research in 1991, our objective was to study the patterns of schistosomiasis transmission in two unspecified village communities in Munufiya governorate, in the central Delta. Building on our experience of water and sanitation-related behaviors and facilities in other Delta villages, our team decided to identify one village that had a piped water supply and one that had this as well as a pipe-borne sewerage system.

As we found that most of the larger villages (*qura*) in Munufiya were served by a pipe-borne water system, but only two villages had a sewerage system, the village with the sewerage system was identified first (al-Garda). Then, matching this community in size and general social structure, we identified a village that only had a water supply (al-Salamuniya); see Appendix A for criteria for village selection. These two communities were the setting for an expanded study that lasted until the end of 1996.

In 1994, we were asked by the Schistosomiasis Research Project to monitor activities associated with a laundry intervention which was planned in a village, also in Munufiya governorate, which we will call Baguriya (all three village names are pseudonyms). In Baguriya, our team carried out baseline and epidemiological surveys similar, but less extensive, than those we had conducted in al-Garda and al-Salamuniya. We also carried out studies of water contact activities. Most of the present study, however, is about the two principal villages, al-Garda and al-Salamuniya.

Establishing a research base

In 1991 members of our team, together with staff from the Department of Health in the governorate and the district (*markaz*), made a number of familiarization visits to al-Garda and al-Salamuniya and met with local leaders, the imams of the mosques, and schoolteachers. In these meetings, team members and MOH staff explained what we were hoping to do, and asked the village leaders to act as partners in the project.

Realizing that we needed a base in each village, the MOH gave us permission to use a room in the al-Salamuniya Rural Health Unit, with an adjacent large veranda. As there was no RHU in al-Garda, the project rented an apartment there to serve as our base. We then identified nursing staff and technicians from the health unit responsible for al-Garda, who then worked with us in our rented premises in that community. Both these research bases were supplied with office and laboratory equipment, including a new binocular self-illuminating microscope which technicians used to identify schistosome eggs in urine and feces.

Drawing on the human resources available in each of the study communities, we hired part-time interviewers and field workers, reputable young people who were already well known in their communities. At any one time, their number ranged from six to ten. Many were young women who had graduated from a university with a degree in social work or a related field and who were waiting for their first full-time job. Other part-time assistants were male university graduates. Their services were particularly useful when it came to interacting with male heads of household and with male employees of health units and Village Councils. Among other local partners were two primary school headmasters, one from al-Garda and the other from al-Salamuniya, who helped with studies of school based screening and health education.

Training

As part of the participatory research strategy, we used local people to carry out as many research tasks as possible. By involving them in this way, we were consciously transferring skills, income, and awareness to local people. As preliminary inquiries had shown that local health providers and technicians had less than perfect knowledge about schistosomiasis, our first task was to provide them some general training. At interactive workshops we reminded them that there were two quite different forms of schistosomiasis, each of which required different testing techniques. Following this the local MOH technicians were briefed on the parasitological techniques required for the study and later sent for training to the MOH Center for Field and Applied Research (CFAR) in Qalyubia governorate.

We also provided information on schistosomiasis, and training in interview techniques, for the interviewers and field workers. Their services would be vital a short time later when we carried out our baseline census. We trained them how to organize and conduct focus group discussions, how to discreetly observe water use activities, and how to follow up people who had been treated for schistosomiasis and needed a further test three months later.

As the funding grant supporting these activities was provided through USAID, the rights of all the people who gave us information and participated in the epidemiological tests were safeguarded by US regulations for the protection of human subjects. We held an interactive training session to introduce interviewers and field workers to these concepts and to ensure that they understood their responsibilities toward the villagers with whom they were interacting.

Baseline data: the census

In preparation for our baseline census and initial epidemiological survey, we numbered each house and prepared a map of the built-up area of each village. In December 1991 and January 1992 (at a time when most people normally resident in the village were on hand) we conducted a census of current residents. We enumerated people within their households, recording their age, sex, education, primary occupation, and work in the fields. We also recorded whether the household had facilities such as piped water, sewerage and electricity connections, and latrines (including type, method, and frequency of evacuation of effluent), type of house construction, and household possessions (such as TVs, radios, and washing machines). We entered these data on a computer using the Statistical Package for the Social Sciences (SPSS), and later linked this information to that obtained during our epidemiological studies.

Epidemiological surveys

At the beginning of the project, no reliable information was available to identify schistosomiasis infection rates in the two communities. Therefore the first task of the epidemiology team was to obtain an estimate of the prevalence of *S. mansoni* and *S. haematobium*, in order to identify an appropriately sized sample for the larger survey. A 5% sample of village residents was tested for both types of schistosomiasis (see Appendix B). On the basis of the findings of this survey, the research team, SRP consultants, and MOH staff involved in planning the survey decided that a 15% sample of the population in al-Garda and a 6% sample in al-Salamuniya would result in a sufficient number of positive cases for the main epidemiological study (around one hundred in each village). In view of the very small number who tested positive for *S. haematobium* (<1%), the group decided to limit the study to *S. mansoni*. This would allow us to focus on villagers' and providers' perceptions of *S. mansoni* and its symptoms, and on diagnosis and treatment procedures. It would also give us more time to look at issues associated with fecal contamination.

In addition to the initial epidemiological study conducted in mid-1992, two further surveys, at twelve month intervals, identified incidence (new cases since the last survey) and reinfection rates reflecting current patterns of disease transmission. All individuals found to be infected were treated with praziquantel and followed up after three months to check on the efficacy of the treatment. The procedures in Baguriya were similar to those in the two principal study communities.

The protocols for urine and stool testing, and for quality control, followed those established for the Schistosomiasis Research Project (SRP), which ensured comparability with studies conducted elsewhere in Egypt. A modified Kato-Katz thick smear technique was used for stool examinations and Nucleopore filtration for the examination of urine, providing for a quantitative assessment of egg output, rather than simply a record of the presence or absence of eggs (Cline et al. 2000; Hussein et al. 2000).

Environmental studies

At the beginning of our project, we realized that we needed information about the chemical and bacteriological quality of various water sources in the two communities—from tap water, handpumps, and canals. Chemical analysis provided a scientific assessment of water quality that could be related to the village women's assessment of the suitability of water from the different sources for domestic tasks. Bacteriological analysis identified possible fecal contamination.

Team members from the High Institute of Public Health carried out tests of water quality in the tanks supplying the water to the villages, and from handpumps. They tested canal water four times a year, a total of eight times between winter 1992–93 and winter 1993–94, at sites in the built-up area of the villages where water contact behavior had been studied, and at canal sites in the fields.

Between April 1992 and December 1993, a malacologist from the High Institute of Public Health in Alexandria supervised the MOH staff who, every month, collected snails in canals in both villages. The snail staff then crushed the snails and examined them under a microscope for schistosomes.

An emphasis on qualitative studies

We used mainly *qualitative* techniques to explore knowledge and attitudes concerning schistosomiasis and its treatment, and gendered social behavior associated with exposure to the disease and the treatment process at the local health facilities. Our findings from these explorations thus went well beyond the numerical information supplied by our baseline census and epidemiological surveys. As researchers, we began by asking: "What do we need to know, and how can we best go about finding it out?" Our procedures required time and conceptual space (see Cline 1995).

For the research team, qualitative research was at the heart of the participatory strategy. An important aspect of our study was the exploration of "process"—what went on at the side of the canal, during a visit to a rural health center, or during an education workshop, and what these activities meant to the various actors involved. The actors assigned meanings to these processes, and expectations concerning what ought to happen. Meanings and expectations were often based on gender roles. These processes could not be expressed in numerical terms, but had to be teased out during our detailed discussions with villagers and local government and health staff working in the village.

We cross-checked the information we collected using a process known as "triangulation" (just as location readings were checked in surveying). For example, in analyzing villagers' water contact activities we compared reported behavior collected during the baseline census to findings from observation studies and in-depth household discussions. Observations could identify what happened at the canal-side, and the in-depth discussions helped to understand *why* people went to the canal. However, different strategies for collecting information about the same topic also provided very different angles on the subject. For example, discussions with household members gave them the opportunity to define their own categories and activities rather than to fit their replies to interviewers' predetermined categories.

Interviews and group discussions

As the project unfolded, we used a number of individual and group discussion methods. We drew up a number of short surveys using open-ended questions that encouraged respondents to express their own ideas freely. For example, we asked those who tested positive during the first epidemiological survey to identify the places where they had contact with canal water, and to say what they had been doing there. Then too, we administered a survey in 80 households, consisting of 226 adults, asking various householders about their knowledge of schistosomiasis, its severity and treatment, and why they thought people persisted in using canals knowing that in doing so they risked schistosomiasis. These surveys helped us to plan our observational studies at the canals and further, in-depth discussions.

We interviewed key informants such as the staff at the governorate, district, and local levels responsible for various aspects of schistosomiasis control. Following Coreil (1995), we were well aware of the difference between an "interview"—in which information tended to flow one way, from the person being interviewed to the interviewer—and a "discussion." Whenever possible, we preferred to spark off a discussion. We thus held group discussions in households, involving everyone, especially women. In these discussions, we focused on the socialization of infants and young children, to find out how they learned about acceptable behavior at the canals.

At various stages in the research, we used focus group discussions to explore a range of conceptual issues. A specialist in the training of focus group discussion leaders trained six assistants to act as moderators and notetakers. The trainer, trainees, and members of the research group then joined in identifying the topics which would be discussed in the focus groups, the settings most appropriate for these discussions, and who should be asked to join them. For instance, we decided to discuss a topic with a group of girls and women, and then with a group of men, in order to identify gender differences in beliefs and practices related to canal water (see Appendix C).

In addition to structured discussions, all members of the researcher team and field assistants kept records of their interactions with all partners in the village and exchanged information on a regular basis. From these exchanges, plans for more structured inquiries often developed.

Observations

We carried out observations of people's behavior at canal sites in order to find out what actually happened there, and compared this information with what local people said they did. Researchers who have conducted observational studies recognize that there is no "best" way to conduct water contact studies, and that they should be designed to answer specific research questions (Blumenthal 1989; Bundy and Blumenthal 1990). Thus, we planned our observations to reflect our objective: to find out what was being done, who was doing it, how it was done, and the parts of the body that were immersed. Some of the issues that emerged from this analysis were taken up in later household group discussions, in order to try to determine *why* a given behavior had occurred. This combination of observations and follow-up discussions is unusual as water contact studies do not usually go beyond identifying and characterizing behavior to ask *why* it occurs.

Identifying possible control strategies

"Action research" has two parts, the initial formative research (such as that we have just described) and the next phase, the "action," which is usually an activity that has been designed and tested in the research setting. When we identified such "action" we focused on activities that seemed feasible and appropriate in the local setting, and did not require any extra staff, money, or equipment.

For example (as will be discussed in more detail in chapter 11) we found that rural health unit staff and villagers were basing their behavior and expectations about testing and treatment on the assumption that *S. haematobium* was the main type of schistosomiasis in the study villages. However, our pilot epidemiological study confirmed that *S. mansoni* was actually the main form of the disease locally, as elsewhere in the Delta. Thus, we had to alert the health staff to this fact and help them to adjust their thinking accordingly. We worked together with the staff at the al-Salamuniya RHU to develop gender sensitive strategies for collecting stool specimens to identify *S. mansoni* during school-based screening. We also designed and tested a modified recording system to help staff follow up with villagers after they had been treated.

As a contribution to prevention activities, we worked with schoolteachers in al-Salamuniya to identify ways to introduce health education about schistosomiasis into regular school programs as well as summer clubs. We also explored other community-based approaches. As mentioned earlier, in Baguriya we monitored the planning stages of the introduction of a laundry for village women, and developed a protocol to encourage women's participation.

In both al-Garda and al-Salamuniya, toward the end of the project the research team and MOH staff shared their findings with community members at village meetings. We initiated a discussion with villagers about possible preventive strategies for disease control, especially ways to limit their exposure to infected canal water. Further meetings were held in al-Salamuniya, in association with the Community Development Association. Based on our environmental studies of water quality, canal ecology, and the distribution of snails in the canals, researchers, village participants, and irrigation consultants explored options for modifying the flow of canals, and getting rid of snails.

The community-based and environmental activities involved collaboration with other ministries, in addition to the Ministry of Health, and were characterized by a preventive approach. All findings from studies in rural health units and schools, as well as environmental conditions and interventions, were presented to the Ministry of Health. In the next chapter we will look more closely at the setting of our study communities.

7

Schistosomiasis in the Village Setting

The village and its farmland

This chapter describes two Nile Delta village communities which, for the purposes of this study, we have given the fictional names of al-Garda and al-Salamuniya. We will begin by identifying the main characteristics of the two communities, as identified by our team in mapping and familiarization procedures, and in a baseline census carried out in late 1991 and early 1992. We will then present the findings of our epidemiological survey.

The two communities are situated in the central Nile delta, in Munufiya governorate, about 70 kilometers north of Cairo. They are large, compact settlements, surrounded by irrigated farmland. When the research project began, al-Garda had a population of around 7,600 and al-Salamuniya a population of just over 8,000. Like many other "villages" that can be briefly glimpsed while traveling between Alexandria and Cairo by rail or along the main "agricultural road," they look like small towns, with closely packed houses and lots of activity in the streets.

The description of these two settlements as *qura*, villages, originated in an earlier time, when most of the population were farmers, and the villages were much smaller than they were by the early 1990s. But, like villages elsewhere in Egypt, they are changing rapidly, in social and economic structure. Villages are differentiated from the urban center that is the headquarters of the district *(markaz)*, as well as from small hamlets (*'izab*, sing. *'izba*) in the surrounding fields. Since the administrative reforms of 1960, both of our study villages have been designated as sub-villages, *qaria tab'a*, but they each have elected representatives on the Popular Village Council in the nearby mother village, the *qaria umm*. They also have some local offices, such as the food supply cooperative for the distribution of subsidized food. Village characteristics are shown in table 7.1.

Physically the two villages are similar in general layout. In the central, older part there are no paved streets, and vehicles negotiate with difficulty between pedestrians, women vegetable sellers, chickens, goats, donkeys, and animal-drawn carts. In al-Garda, the main mosque and the houses of the appointed village head (the *'umda)*, are in a small central square. In al-Salamuniya, the main mosque and the *'umda*'s house face the main canal that threads its way through the center of the village.

In the heart of both communities, the narrow alleys that serve as thoroughfares are lined with one and two story houses, made of mud brick and often plastered and lime-washed. These old-style houses are built around a *hosh* or courtyard, which is entered through a double door from the street.

Table 7.1: Characteristics of al-Garda and al-Salamuniya, 1991–92

Village facilities	al-Garda	al-Salamuniya
Health Unit	no	yes
Private doctors offices	—	2
Private doctors	4	15
Village Council	no	no
Schools		
Primary schools	3	2
Preparatory schools	1	1
Food supply cooperative	1	1
Post office/telecom center	1	1
Agricultural cooperative	1	—
Library	2	—
Youth center	—	1
Community Development Association (+ clinic)	—	1
Market	weekly	twice weekly

Source: village survey, December 1991

In both communities, moving out beyond the village center one finds modern style housing, built of fired brick and concrete. Many of these houses, following the city fashion in plan, contain a number of separate, self-contained apartments. Each has its own separate entrance from the stairwell, and its own kitchen and bathroom. Such buildings enable members of extended families to live in close proximity. The living space can gradually expand upward by adding a new apartment as each son marries and brings his bride home. Even if a son goes to work in a nearby town, or in Cairo or the Gulf, he still has a place to return to in the village.

All village houses, whether made of mud brick or of fired brick and concrete, have a flat roof. These are used for hanging out washing, drying fodder and grain, keeping pigeons and chickens, and as a place for family sociability. These features reflect the rural character of the community.

In al-Garda and al-Salamuniya, as elsewhere in rural Egypt, the spatial structure of the village reflects the dual nature of the local power structure. The government facilities which represent central authority—the post and telephone office, the agricultural and food supply cooperatives, schools, and (in al-Salamuniya) the Rural Health Unit—are on the edge of the village, adjacent to the main road and its access to the urban world, represented by Cairo. But lying near the center of each of the tightly packed villages are the buildings associated with the "traditional" power structure—the main mosque, the meeting house *(mandara)*, and the houses of the *'umda* (the village head) and the heads of the leading families.

Local government primary health care facilities in many Egyptian villages provide diagnosis and treatment for schistosomiasis. Al-Salamuniya has a Rural Health Unit (*wihda sihhiya)*, built in 1962. Residents in al-Garda must go to the rural health unit in the nearby mother village one kilometer away. Al-Salamuniya also has a small clinic operated by the Community Development Association (the CDA is in part supported by government and in part by NGOs). In both villages, there are also resident private doctors who visit patients in their homes and, in al-Salamuniya, private doctors who have their own offices. However, there are fewer private doctors in al-Garda than in al-Salamuniya, and no NGO clinics, as the village is only three kilometers from the governorate center of Shebbin al-Kom. Both villages have small pharmacies where medicines and medical supplies can be purchased.

Canals and drains flow through the built-up areas in al-Garda and al-Salamuniya (as in many other Egyptian villages). These provide many sites that young girls and women can use for domestic activities, as we will explain later in greater length. Some sites are used by men for washing vehicles and animals; others are ideal swimming places for adolescent boys.

Figure 7.1: The village of al-Garda

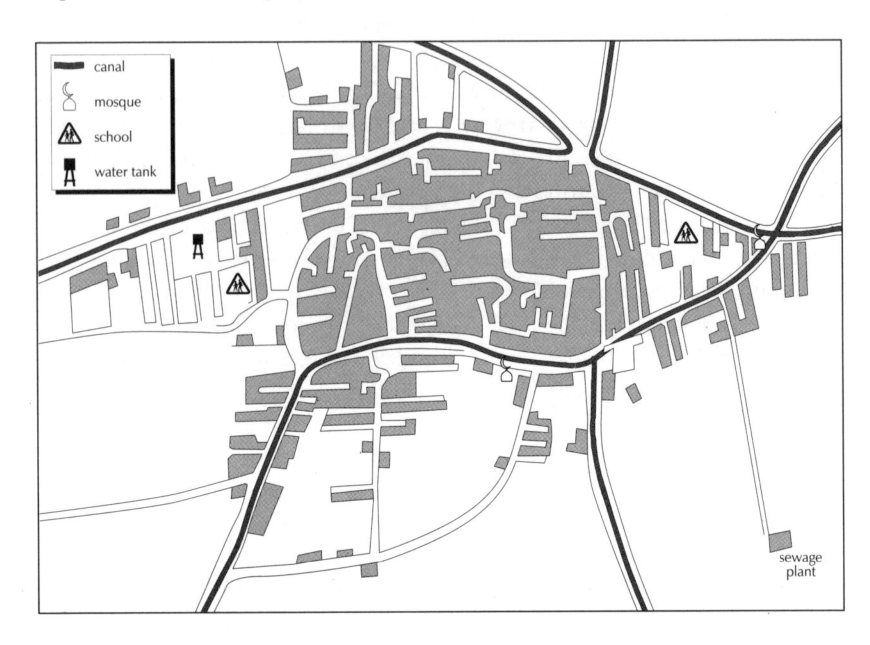

Figure 7.2: The village of al-Salamuniya

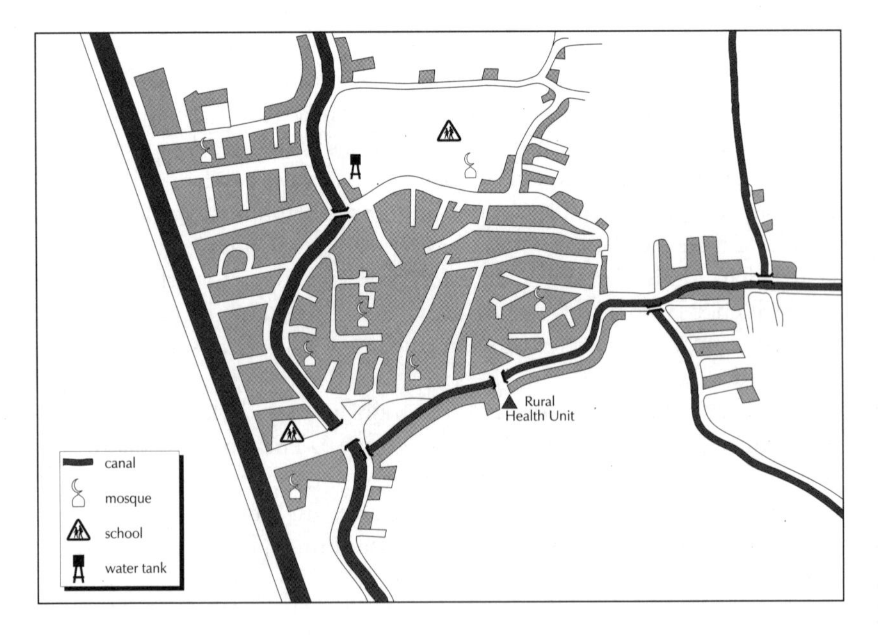

Both al-Garda and al-Salamuniya are surrounded by densely cultivated, irrigated land, the village *zimam*. It consists of small plots, some no more than twenty or thirty meters wide, which are used for growing vegetables (especially *mulukhiya*, a green spinach-like vegetable popular for soup), wheat, maize, cotton, and *berseem* (the fodder crop, alfalfa). The *zimam* of al-Salamuniya comprises 1,600 feddans (1 feddan = 1.038 acres or 0.42 hectares); that of al-Garda is much smaller, only 700 feddans, reflecting its position in a more densely populated rural area close to the governorate city.

A few houses and shelters for cattle are found in the *zimam*. In the case of al-Salamuniya, two small *'izab*, farming hamlets, still maintain their independent identity. One is situated just across the main canal from the village and the other is about two hundred meters away, in the fields.

As occasional winter rains are not enough to support farming, cultivation depends entirely on irrigation. The principal canal serving al-Garda is parallel to the main road, along which are located many of the village's commercial activities—shops and transport repair workshops—as well as several sites where women regularly wash clothes and dishes. Al-Salamuniya's principal canal is located on the far side of the main railway and the busy Alexandria-Cairo agricultural road. Thus, it is not a focus for commercial or domestic activity.

Running through the center of both villages are secondary canals that attract local women doing their domestic chores. These canals were built to provide the water for the *misqas*, small field canals that are generally blocked up and dry until the farmers' turn to irrigate their fields comes around. Then they must use a lifting device, a water wheel or diesel pump, to lift the water into the field ditches that supply their land.

The people of the village

The resident population of al-Garda at the time of the baseline census of 1991–92 was 7,677, as shown in table 7.2. Al-Salamuniya was slightly larger, with 8,181 people. In both cases there were approximately twice as many people living there as had been recorded thirty years earlier in the 1960 census, when there were 3,900 people in al-Garda and 4,937 in al-Salamuniya. This doubling of the population is not at all unusual in Egypt. On the national level, population increased from 26 million in 1960, to 50.4 million in 1986, and 59.3 million in 1996. In both villages, males were slightly more numerous than females, as in Egypt as a whole; 51.2% of the population was male in both the 1986 and 1996 censuses.

Table 7.2: Population of al-Garda and al-Salamuniya, 1991–92

	al-Garda		al-Salamuniya	
Male	3945	51.4%	4281	52.3%
Female	3732	48.6%	3900	47.7%
Total	7677		8181	

Source: baseline census, December 1991–January 1992

The population structure of both communities was similar, with relatively few older people, and a large proportion of young people, showing the pyramidal structure common in most low income countries. In al-Garda, 38% of the population was under 15, compared to 42% in al-Salamuniya. In both communities, the proportion of the population over 60 was similar: 6.5% in al-Garda and 5.38% in al-Salamuniya. These figures were not very different from those for Egypt as a whole, with only 5.6% of the population over 60 in the 1986 national census. Our village census identified an approximately equal proportion of older males and females, again, similar to the figures reported in the national census. All in all, these figures suggest a fairly stable population. The absence of marked differences in the number of male and female villagers of the same age suggests relatively little permanent out-migration.

Household structure

Households were larger in al-Garda, with an average size of 6.53 people compared to 5.06 al-Salamuniya. This contrast is reflected in the larger proportion of households in al-Salamuniya (89%) which consisted only of a nuclear family (mother, father and children), than in al-Garda (69%). Thus, households in al-Garda are more likely to include an older parent (or parents) and one or more married sons and their families.

The larger households usually resulted when a newly married son brought his wife home to live with him as part of his parents' household (daughters, in contrast, moved into their husband's family home and were thus "lost" to their natal family). Some larger households consisted of two or more married brothers and their families living with one or both parents. Such a household might split up upon the death of the father or mother, or if one of the more independently-minded sons decided to set up a separate household within the existing building or move to a new house altogether.

In the two study communities, as in Egypt as a whole, marriage is basic to the foundation of a household unit. Very few Egyptians remain unmarried. In the two study villages, for any age group over the age of 40 the number of males or females who had never married was rarely more than 3%. However, the lifetime marriage experience of women and men was rather different. Men usually married later than women, so in al-Garda, among adults aged 25 to 29, 87% of the women were married, compared to only 32% of the men. In contrast, toward the end of the life cycle fewer women than men were married, for women who become widows or are divorced are less likely to remarry than are men. Thus, in al-Garda, just over a quarter of the women in the 50 to 54 age group were widows, compared to 1.5% of men.

In both communities, a few adults, mostly men, lived away from the village, but they were still thought of as belonging to a particular household and were obliged to contribute to its upkeep. Few women left; most accompanied their husbands rather than working independently.

Three times the number of people from al-Garda worked outside Egypt compared to al-Salamuniya; only eight from al-Garda and none of those from al-Salamuniya were women. However, the situation was very different when considering those who left to work within Egypt, beyond the governorate of Munufiya. In this case, nearly three times the number of people left al-Salamuniya, 448, compared to only 174 from al-Garda. This reflected the greater accessibility of al-Salamuniya, on the Cairo-Alexandria agricultural road, and its proximity to the governorate of Gharbia, of which it was once a part.

In both study communities, our baseline census showed that some households were recorded as being headed by women. Most of these households consisted of a widowed woman who did not have an adult son. In a few of the others, the woman was only temporarily acting as head while her husband was working away from the village. Overall, our census recorded that 10% of the households in al-Garda and 13% in al-Salamuniya were headed by women. Although other studies, such as that by Nagi (2001: 61–64), in 1995, have suggested that female headed households were economically disadvantaged (see chapter 2), we have little information that would substantiate this claim in our study villages.

Living conditions

Within each of the study communities, living conditions varied considerably. Modern, well-appointed apartments with full kitchens and bathrooms existed side by side with mud brick houses without an indoor water supply. Overall, living conditions were slightly better in al-Garda, where the proportion of households living in mud brick houses was less than half that in al-Salamuniya, as shown in table 7.3. In mud brick houses, cooking was generally done on a primus stove, whereas in a modern house made of fired brick and concrete, cooking was done on a fully modern stove (with an oven) which was supplied with a cylinder of butagaz.

Although piped water was available in both communities, far more households in al-Garda had individual water connections. In the case of a mud brick house, the water connection usually consisted of a single tap, situated in the courtyard adjacent to the main entrance from the street. In most of the fired brick and concrete houses, a number of indoor taps were found in both the bathroom and cooking areas.

Table 7.3: Household characteristics

	al-Garda	al-Salamuniya
Number of households	1175	1618
HH in mud brick houses	25%	58%
HH with electricity	97%	90%
HH with water connections	78%	39%
HH with toilet/latrines	98%	94%
HH with pipe-borne sewerage	33%	none

Source: baseline census December 1991–January 1992.

In the 1991–92 census, over 90% of all households in the two communities were reported to have their own toilets or latrines, either within the house or in the courtyard. At the time of the baseline survey only one third of al-Garda's households were connected to its sewerage system; however, as we will see in the next chapter, within two years the number of such connections had doubled.

In al-Salamuniya, and in households in al-Garda not connected to the sewerage system, toilets and latrines connected to a closed tank required regular emptying, while in others the effluent seeped gradually into the subsoil. A few latrines and kitchen drains were connected to pipes leading to the canal. From a health point of view, few of these facilities could be regarded as safe.

The majority of households in both villages had electricity, as shown in table 7.4. Electricity provided power for lighting and household appliances. Radio and TV sets were widespread, providing access to news media, as well as to soap operas and public health messages about the do's and don'ts of good health.

Table 7.4: Household appliances, 1991–92

	al-Garda		al-Salamuniya	
Radio	1059	94%	1303	85%
TV	1008	90%	1280	83%
Washing machine	986	88%	1275	83%
Refrigerator	565	50%	541	35%
Sewing machine	158	14%	155	10%
Total households	1122		1541	
missing values	130			

Source: baseline census 1991–92

Many households also had a washing machine, usually a small semi-automatic affair that could do only part of the laundry tasks—soaking and agitating the clothes, but not rinsing them. Some households had a refrigerator, which was often used only in summer, to store water and, to a lesser extent, food. Refrigerators were used year round by small shopkeepers selling soft drinks, ice cream, and snacks.

Earning a living

Contemporary rural Egyptians derive their income from a variety of sources, reflecting the changing structure of rural society and the growth of education and urban type employment opportunities within the village since the 1952 Revolution. By the 1990s it was estimated that only half of rural income was obtained from agriculture (Hopkins 1999:369). In this respect, al-Garda and al-Salamuniya were not atypical.

Although the villages were situated among irrigated fields, our 1991–92 census showed that less than one quarter of the males over 15 in each community were full-time farmers, as shown in table 7.5. Less than 1% of women said they were full-time farmers. However, in themselves, these figures are grossly misleading since they conceal the fact that a high proportion of the people in both communities spend part of their time cultivating plots in the *zimam*. This part-time farming in irrigated fields, no less than full-time farming, puts them at risk for schistosomiasis.

When asked directly about their agricultural activities, a third of the women in our two communities reported that they spent some time each day working in the village *zimam*: 38% in al-Garda and 33% in al-Salamuniya. Not surprisingly, a further quarter of the adult males (in addition to the full-time farmers) reported that they farmed every day, after returning home from their regular job and on Friday (their day off). Similarly, many school-aged children also worked in the fields, especially during the long summer vacation. Another category of village people working in the *zimam* were young people who had completed their university education and were waiting to be assigned a government job. Given the general economic situation in the early 1990s, this "waiting" period might last for several years.

In the two communities as a whole, most women were reported as "housewives," *rabit manzil*, as shown in table 7.5. If one excludes the number of young women over 15 who were still students at the secondary and tertiary level, the proportion of those who were "housewives" increases to 86% in al-Garda and 90% in al-Salamuniya. As we mentioned in chapter 2, the general use of the term "housewife" reflects the belief, among both women and men, that a woman's primary responsibility is to her family. However, many women who called themselves housewives worked in the fields or in the village selling agricultural produce.

Table 7.5: Occupational structure of those 15 years old and above

	al-Garda		al-Salamuniya	
	M	F	M	F
Formal sector, primary plus	537	160	515	
	21.8%	7%	20.5%	88.4%
Farming	566	14	568	14
	23%	<1%	22.5%	<1%
Skilled and unskilled workers,	682		617	12
service workers, shop keepers etc.	27.7%	45.2%	24.5%	<1%
Students, military	403	354	509	285
	16.4%	15.4%	20.2%	12.8%
Housewives, or not working	127	1631	110	1719
due to illness, age	5.2%	70.9%	4.4%	77.3%
Unemployed	145	95	205	107
	5.9%	4.1%	8.1%	4.8%
Total	2460	2299	2524	2225

Source: baseline census, 1991–92

In our 1991–92 baseline census, around half of the adult males in the two communities were employed in the informal sector. Half of these were full-time farmers, and the rest were skilled and unskilled workers in trades and handicrafts; a few owned shops and small businesses. Around one fifth of the males worked in formal sector jobs that required at least a primary school education, mostly as *muwazzafin*, government employees. Higher level posts required a university education. In al-Garda, 5.2% of the men and 1.3% of the women were employed in these university level jobs, compared to 6.9% and 1% respectively in al-Salamuniya.

In both communities, many of the young men were serving in the military or were still acquiring a formal education. Others in this age group were technically unemployed, but claimed to be "waiting" for employment. The educational reforms of the 1950s guaranteed all qualified young people access to a university education and a government job after graduation. While today education is still free, there is no longer a guarantee of a government job immediately after graduation—the waiting period might be five years or more. The extent of graduate unemployment is revealed by the fact that in al-Garda, 14% of male university graduates and 13% of women graduates were recorded as unemployed, or as they preferred to put it, "waiting"; the comparable figures for al-Salamuniya were 17% and 41%.

Education

In the broadest sense of the word, it is clear that education can play an important role in the control of schistosomiasis. At the time of our study, al-Garda had four schools. Three of these were primary schools, with a total enrollment of 1,412 children. The fourth was a preparatory school, with 585 children. Because of a shortage of classroom space, two of the primary schools operated on a double shift in the morning. The preparatory school operated in the same building on an afternoon shift. In al-Salamuniya there were about 1,500 children in two primary schools, and 600 in the one preparatory school. Here, the two primary schools used the same rooms, one for the morning shift and the other during the afternoon.

Because neither community had a secondary school for those aged 15 and over, all young people going beyond the compulsory preparatory level had to travel daily (usually by minibus) to a secondary school in a nearby village. All students going on to tertiary education had travel even further outside the village, although, as we have seen, many returned later to "wait" for government employment.

In looking at education and literacy in the two communities, as in the rest of Egypt, two features stand out; an overall increase in educational levels over time, and the gradual decrease in the education gender gap. First, looking at the proportion of the population above the age of compulsory schooling, our census showed that twice as many women as men were functionally illiterate. In al-Garda, 57% of the women and 27.3% of the men were unable to read or write, and in al-Salamuniya 68% of the women and 32.8% of the men were similarly handicapped.

Young people were much more likely to have received an education than were their parents, and among them the gap between schooling for girls and boys was smaller. The striking feature about education in the two communities is the high proportion of primary and preparatory school aged children, especially girls, currently attending school. As we shall see later, their presence in school is especially important as a major aspect of the Ministry of Health schistosomiasis control activities is the school based screening and treatment of children. Children not attending school do not have access to this service. As shown in table 7.6, over 90% of boys of school age and a slightly smaller proportion of girls regularly attended school. We rechecked the results of our census, in case parents had been tempted to report that their children attended school when they did not (schooling is compulsory up to the age of 14, usually reckoned as the end of the preparatory school). But the teachers in both villages, who worked closely with our research group, confirmed these high attendance rates.

Table 7.6: Current school attendance for those aged 6 to 14

	al-Garda		al-Salamuniya	
Boys:				
Not in school	41	(4.7%)	63	(6.3%)
In school	816	(95.3%)	932	(93.7%)
Girls:				
Not in school	93	(11%)	133	(14%)
In school	735	(89%)	816	(86%)

Source: baseline census 1991–92

However, if one looks at the proportion of girls *not* in school, more than double that for boys, the picture looks rather different. Even in a relatively favored area of rural Egypt, such as the governorate of Munufiya, the discrepancy between the school attendance of girls and boys is considerable. In short, gender differences continue to impact heavily on a young person's future, including, as we shall see later, their chances to obtain diagnosis and treatment for schistosomiasis in a school-based program.

In our two study villages, an increasing number of young men and women have received, or are receiving, a university education. There were many more graduates under the age of thirty than above, and the diminishing gap between the number of men and women graduates reflects a more equal access to higher education. In al-Garda 35% of the graduates under thirty were women, and 28% in al-Salamuniya.

Schistosomiasis in al-Garda and al-Salamuniya

In mid-1992 our research team conducted an epidemiological survey in al-Garda and al-Salamuniya in order to identify the proportion of those infected (for sampling details see the previous chapter). We matched these findings with those from our 1991–92 baseline census data in order to identify the characteristics of this sample population.

Perhaps the most important finding of our 1992 epidemiological survey was that the prevalence of *S. mansoni* in al-Garda (8%) was one third of that in al-Salamuniya (25%), as shown in table 7.7. Contemporary surveys being taken elsewhere in Egypt by EPI 1, 2, 3 also showed that it was not unusual for villages in the same governorate to have widely different rates of infection (see chapter 5).

We chose our study communities on the basis of water and sanitation facilities, with al-Salamuniya only having a water supply and al-Garda having both water and a sewerage system. In all other respects they were similar. At first it seemed significant that the community that had the higher rate of water connections and a sewerage system (al-Garda) had an infection rate that was only *one third* that of the other community (al-Salamuniya). However, a closer look at the data showed that there was no clear correlation between residence in a house without a water connection and the likelihood of an individual in that household being infected. We found that, although in al-Salamuniya a significantly higher proportion of infections occurred in households without water connections, this was not the case in al-Garda. We also failed to identify a correlation between infection and residence in a mud-brick or modern house.

In al-Garda there was actually a larger percentage of infections in households that *were* linked to the sewerage system, compared to households without connections (Watts and El Katsha 1997). In the chapter following, on environmental conditions, we will explore the possible impact of water and sanitation on infection at the individual and community level in more detail, as well as looking at the possible impact of other environmental aspects, such as canal characteristics and snail populations.

As in most other communities studied in Egypt, in both our villages overall infection rates were higher for males than for females, as shown in table 7.7

and figures 7.3 and 7.4. The gender differences in both villages were highly significant. The difference in al-Garda was greater, with almost double the number of males than females infected, whereas the female infection rate in al-Salamuniya was two thirds that of the males.

Table 7.7: Prevalence of *S. mansoni*, 1992

	Males			**Females**		
	# tested	Positive	%	# tested	Positive	%
al-Garda						
<5	75	—	—	58	—	—
5–14	156	7	4.5%	140	8	5.7%
15–24	74	12	16.2	94	9	9.6
25–34	36	8	22.2	82	5	6.0
35–44	57	11	19.3	59	3	5.0
45–54	41	5	12.2	29	1	3.4
55–64	20	6	30	24	1	4.2
>65	25	3	12	22	1	4.5
Total	484	52	10.7%	508	28	5.5%
		al-Garda total: 992				
al-Salamuniya						
<5	43	2	4.8%	29	—	—
5–14	60	12	19.7%	85	10	11.8
15–24	30	11	36.6	32	12	37.5
25–34	31	14	45.2	37	14	37.8
35–44	27	11	40.7	24	6	25.0
45–54	8	4	50.0	13	3	23.0
55–64	12	7	58.3	12	1	8.3
>65	5	3	60.0	6	2	33.3
Total	216	64	30.0%	238	48	20.2
		al-Salamuniya total: 454				

In both communities (in line with findings elsewhere) infection rates varied considerably with age. In both villages, the number of boys and girls below five years old who were infected was negligible (2 boys were infected of a total of 205 boys and girls tested). Infection rates rose through the school years and from adolescence to young adulthood. In both villages, the peak in rates among females occurred between the ages of 25 and 34; this was especially notable in al-Salamuniya. Among males, however, the highest infection rates were found among the small number of older males. These infection patterns suggested the importance of gender (reflected in gendered occupational and activity structure, with age reflecting different occupational patterns during the life cycle) in influencing infection rates.

Prevalence rates among full-time male farmers were higher than for adults in any other occupation. In al-Salamuniya, with 15 of 27 farmers (55%) infected, it was significantly higher than for any other occupational group; in al-Garda, although the rate was higher than for other groups, the relationship was not statistically significant. No correlation between infection and other occupations or with educational level could be identified in either village. In chapter 9 we will present more detailed information on water contact that will help us to interpret these findings.

Figure 7.3: al-Garda prevalence of *S. mansoni*

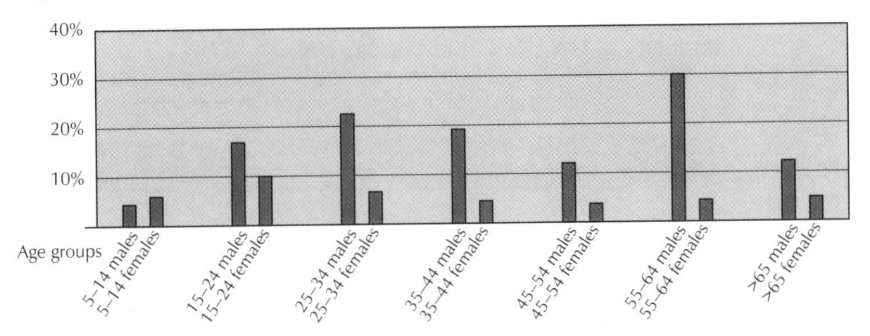

Figure 7.4: al-Salamuniya prevalence of *S. mansoni*

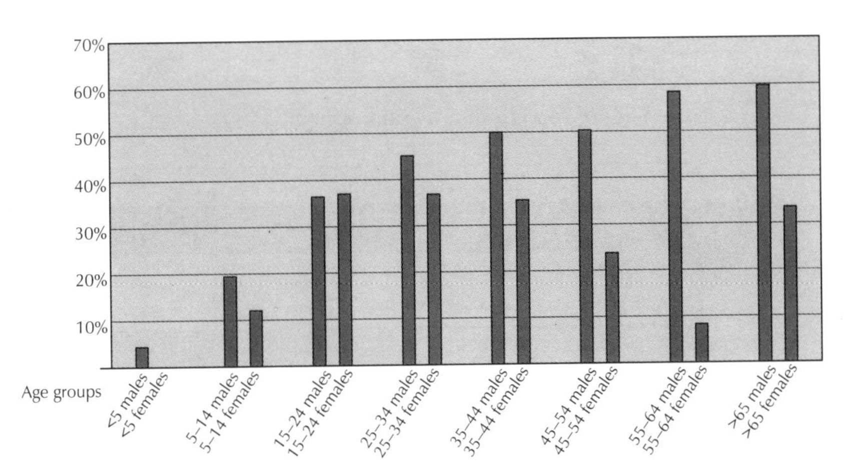

Overall, in the two study communities the intensity of infection, measured by egg count, was low. The majority of those infected—72% in al-Garda and 69% in al-Salamuniya—had low egg counts (less than 100 eggs per gram of feces). Only 11% of those infected had high egg counts (200 or more eggs per gram of feces), with a significantly higher proportion in al-Salamuniya (13.4%) compared to al-Garda (8.7%). In al-Salamuniya the proportion of males and females with high, low and medium egg counts was similar, but in al-Garda females had lower egg counts than males, as shown in figure 7.5.

In line with epidemiological findings elsewhere, young people aged 5 to 14 showed the highest overall egg counts, with 16% of individuals having a high intensity, compared to 10% for those 15 years of age and above.

Figure 7.5: Intensity of infection in al-Garda and al-Salamuniya, 1992

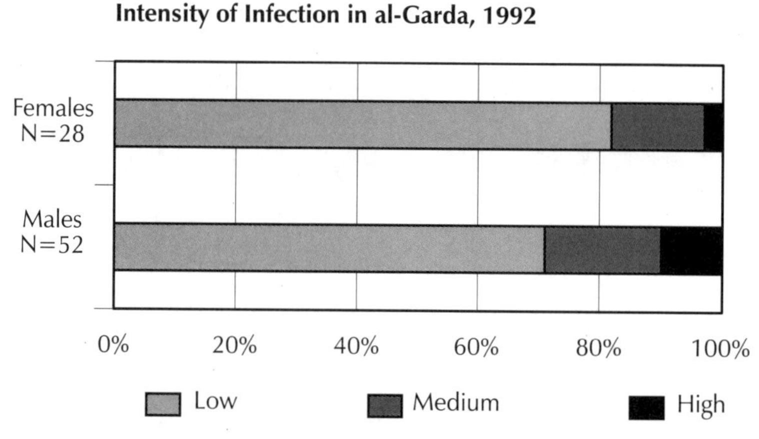

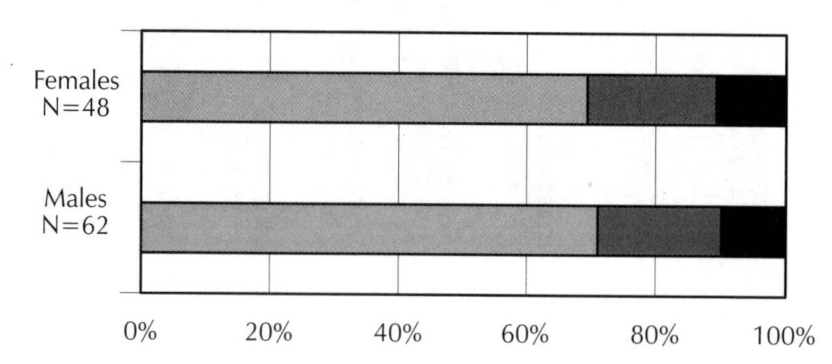

Three months after our initial epidemiological survey, 15% (25) of the 165 positive individuals who had been treated again tested positive. This was due to either treatment failure, which is likely to occur in 20–40% of *S. mansoni* cases (El Khoby et al. 2001), or to re-infection since treatment.

Our second epidemiological survey, conducted in mid-1993, a year after the first, showed that in both communities transmission was still occurring. In al-Garda, the incidence rate (new infections during the year) was 3.4%, compared to 6.5% in al-Salamuniya. These figures reflect the difference in the initial prevalence in al-Garda (8%) and in al-Salamuniya (25%).

The household and spatial clustering of infection

Researchers have long recognized that some cases of infectious diseases are clustered in households, and that this clustering reflects common patterns of disease transmission based primarily on physical proximity. Among the diseases identified with clustering are infections that are environmentally transmitted (through contact with infected water or fecal matter), such as schistosomiasis, ascariasis, and Guinea worm (Cairncross et al. 1996). This clustering can also be understood as the result of households' shared patterns of behavior, such as hygiene practices involving water use and exposure during water contact.

In our epidemiological survey, we were alerted to the possible occurrence of clustering in households when we noted that 16 of 80 infected people in al-Garda (20%) and 29 of 112 in al-Salamuniya (26%) lived in houses in which two or more people were infected; in al-Salamuniya six households had four or more cases. Using a simple test for significance, with expected values calculated from a binomial distribution, we found that schistosomiasis in both villages was indeed clustered in households, as shown in table 7.8. Clustering was more marked in al-Salamuniya, with a higher overall prevalence, than in al-Garda.

Although this simple analysis does not take account of the size of the household, or differences in the age or gender of the household members, it does suggest the importance of exposure behavior that originates in the household, and patterns that are learned and maintained within the household unit. These issues are followed up later in the book.

We also considered the possibility that spatial clustering of the places of residence of infected people might reflect household members' decision to use a nearby infected canal (see figures 7.6 and 7.7). However, we discarded this hypothesis. Even if such spatial clustering did exist it was not likely to explain the distribution of schistosomiasis infection in most instances. For example, in al-Salamuniya, we found that the stretch of canal near the Rural Health Unit was highly polluted, often stagnant, fly-ridden and smelled badly; accordingly it was never used, even by those people living in the six residences along the canal in which there were infected people.

Table 7.8: Household clustering of infection, al-Garda and al-Salamuniya, 1992

1. No of cases	2. No. of HH Observed	3. Col (1) x Col (2)	4. No. of HH expected	5. $(E-O)^2/E$
al-Garda				
0	110	0	104	0.31
1	42	42	50	1.19
2 or more	16	38	12	1.47
Total	168	80		
Mean	0.48			
Chi-square				2.97
al-Salamuniya				
0	40	0	33	1.67
1	31	31	37	0.84
2	16	32	20	0.97
3	7	21	8	0.05
4 or more	6	28	2	6.9
Total	100	112		
Mean	1.12			
Chi-square				10.50

During our later investigations, discussed in chapter 9, we found that women did not necessarily visit the washing site nearest to their house, and many regularly used more than one site. We also found that farmers (the most highly infected occupational group) were likely to have been infected in the water courses flowing by their fields.

The epidemiological findings we have presented in this chapter are useful background information. However, they do not in themselves lead to any understanding of *why* there was such a marked difference in infection levels, with the al-Garda level one third that of al-Salamuniya. In the next chapter we approach this problem by exploring contrasts and similarities in the environmental conditions found in our two study communities that could be associated with schistosomiasis transmission.

Figure 7.6: Distribution of places of residence according to infection, 1992 al-Garda

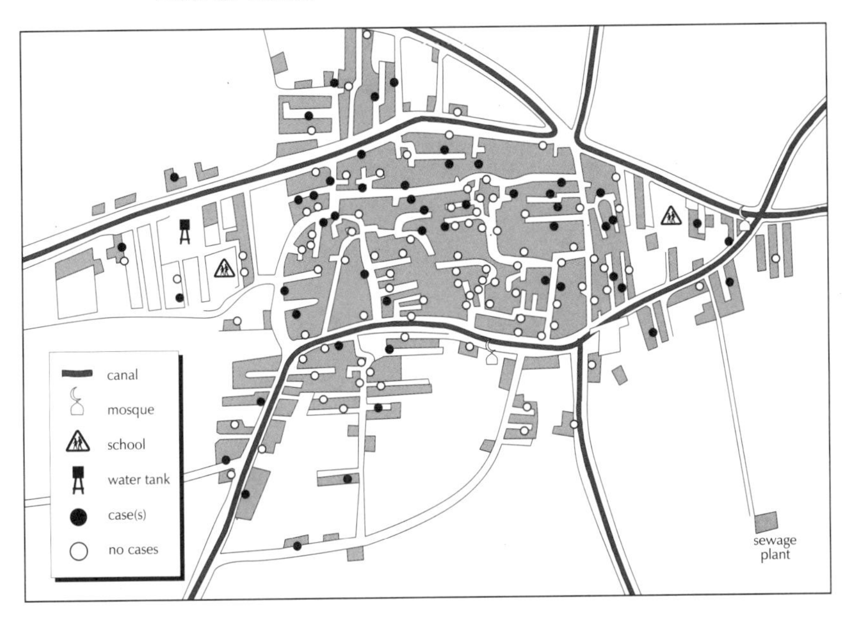

Figure 7.7: Distribution of places of residence according to infection, 1992 al-Salamuniya

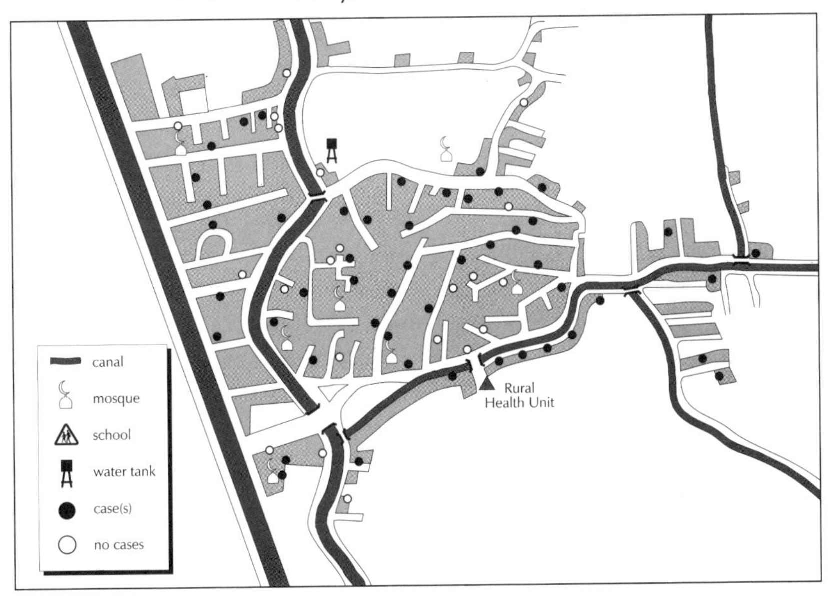

8

The Environmental Setting

The environmental problem

When we began our research project, we selected two study villages. Both had a piped water system and one had, in addition, a pipe-borne sewerage system. As we saw in the previous chapter, apart from this the villages were similar. Both had experienced rapid population expansion in the last half of the twentieth century, growing to around 8,000 people by 1992. Both contained a similar mix of occupations, and housing types.

Yet, in al-Salamuniya, in the community that did *not* have a sewerage system and had a lower proportion of household water connections, the prevalence of schistosomiasis was three times that of al-Garda. This simple fact gave rise to our initial premise that there was possibly some kind of a correlation between high schistosomiasis infection and the lack of individual water connections and a sewerage system that disposed safely of human waste (containing among other things, *S. mansoni* eggs).

In this chapter we will first of all look at the epidemiological data relevant to our initial premise. We will then go on to explore the broader relationship between infection and environmental conditions, comparing the water-related system found in the two communities, and looking at the various ways they may be linked to the transmission of schistosomiasis. Finally, and perhaps most important, we will examine the canal systems in each community that support the vector snails and the schistosome disease agents. It is in the canals that humans are both exposed to the disease and continue the cycle through contamination with fecal matter containing schistosome eggs.

In the early 1990s, a technical explanation for the environmental problems found in al-Garda and al-Salamuniya (and in many other Delta communities) would start with the high groundwater table that caused a ponding of polluted water in low-lying areas and the flooding of sewage vaults. The high water table was the result of a number of interrelated factors. These included the rapid increase in the population and in the amount of water used. They very often also included the lack of effective canal irrigation drainage and the lack of a sewerage system to bring wastewater and human wastes to places where they could be properly treated. In some cases, poorly maintained water pipes and sewerage vaults were also factors.

In the two villages, as in the Delta as a whole, there has rarely been adequate provision for sanitation, or for household drains to cope with the water brought in by a piped water system. In Egypt between 1980 and 1991, the United States Agency for International Development (USAID) Local Development projects devoted 35% of their block grants for rural infrastructure to extending piped water supplies in rural areas; by 1992 this had involved a total expenditure of US $125 million. But rather than funding a regular drainage/sewerage program, USAID supported pilot projects testing the suitability of relatively sophisticated and expensive wastewater and sewage treatment technologies. However, nothing came of these projects. Meanwhile, Egyptian canals continued to be flooded with contaminated water and villagers suffered from fecally contaminated ponds and overflowing sewage vaults (Watts and El Katsha 1995; Gemmell 1989; White and White 1986).

The lack of a proper system for the removal of domestic wastewater was exacerbated by the increase in the level of the irrigation water in canals since the completion of the Aswan High Dam in 1964. The problem originated in the failure of British engineers to install an effective drainage system after the expansion of the irrigation system that began in the 1880s. A century later, the problem was only beginning to be addressed (Ruf 1995).

In the two study villages, the absence of a garbage collection system also contributed to the pollution of the village and the canals. Here, as elsewhere in large Delta villages, more people were producing more rubbish, much of which was non-biodegradable (such as tin cans and plastic bags), and few people still used a mud brick oven (*furn*) which could be kept going with any kind of combustible waste material.

The impact of water and sanitation systems on infecton

Public health experts have long recognized the important role played by the provision of safe water and sanitation in facilitating good hygiene practices and promoting good health. Epidemiological studies have shown that such interventions can help to prevent infections such as childhood diarrhea and trachoma (a potentially blinding eye infection), as well as schistosomiasis (Esry 1996; Boot and Cairncross 1993: ch. 2; WHO 1993: 1; Jordan et al. 1992; Esry et al. 1991; Noda et al. 1988; Feachem 1984). In the Nile delta, the recent increase in cases of lymphatic filariasis (a parasitic infection causing elephantiasis) has been related to the increase in breeding sites for the insect vector, *Culex pipiens*, in heavily polluted stagnant pools and puddles (Harb et al. 1993). Indeed, the need to explore the relationship between schistosomiasis infection and water and sanitation provision was the reason we identified one village with only a piped water system, and another that also had a sewerage system.

Turning first to the potential impact of domestic water supplies on infection, in al-Salamuniya our 1991–92 epidemiological survey showed that there were significantly more infections in households without connections. However, this was not the case in al-Garda, as shown in table 8.1.

Table 8.1: Infection with *S. mansoni* and household water connections

Prevalence, 1991–92

	With connection			Without connection		
	No. tested	**Positive**	**%**	**No. tested**	**Positive**	**%**
al-Garda	772	61	8.4%	220	19	8.6%
al-Salamuniya*	195	31	15.9%	259	81	31.3%

*Pearson Chi-square, p = 0.00017

Why did having individual water connections apparently protect against infection in the community with the smaller proportion of connections (al-Salamuniya) but not in the community with the larger proportion of connections (al-Garda)? What other factors needed to be taken into consideration? As they stand, these findings suggest a complex relationship between access to piped water supplies, contact with canal water, and infection. Our colleague Amal Khairy hinted at this in her study of small villages south of Alexandria, some without any access to potable water and others with access to standpipes. Her study suggested that water *quality* as well as *availability* affected patterns of water use, and hence of schistosomiasis infection (Khairy et al. 1986).

Turning now to consider the possible impact of a system to remove wastewater as well as sewage, we will look at the situation in al-Garda. The lower level of infection in al-Garda does, initially, suggest that such a system might be responsible, in part, for the lower *overall* level of infection in that community. In so far as an effective system to remove wastewater from the house would result in fewer women visiting canals, one could also hypothesize that such a system would result in lower infection rates among *individuals* in served households. However, only one third of the households in al-Garda had access to the sewerage system at the time we conducted our 1991–92 baseline census. In the epidemiological survey in mid-1992, we found that individuals living in houses *with* a sewerage connection were slightly *more likely* to be infected with *S. mansoni* than were people living in a house *without* benefit of sewerage. Nine percent (55 of 610) were infected in households with a sewerage system, compared to 6.5% (25 of 382) living in households without a connection. Thus, our epidemiological data did not suggest that, when considering schistosomiasis infection, such a system directly benefited individuals living in households with a sewerage connection although, as we shall see later, it brought benefits to the whole community. These findings prompted us to explore in greater depth the characteristics of these water and sewerage systems, their use and maintenance, and their benefits to the community as a whole as well as to individual households.

Domestic water supply and water use

As early as 1955, a water system was constructed in al-Garda village that provided public standpipes for its residents and, later, individual water connections. The water system supplied al-Garda and a larger neighboring community; by 1991 it provided for a total population of about 20,000. By way of contrast, the water system in al-Salamuniya was constructed much more recently, in 1988. Largely because this system had been so recently installed, only 39% of al-Salamuniya's households (632 of 1617) had individual connections at the time of our 1991–92 baseline survey. In al-Garda, with its longer established system, 78% of households (914 of 1175) were connected.

Women in al-Garda and al-Salamuniya could choose between several water sources for domestic use. In the past, shared public standpipes had provided an important source of safe water in both villages. The three standpipes in al-Garda had long since been dismantled. In al-Salamuniya, the Village Council closed down the three public standpipes in 1991, claiming that this would conserve water and that individual connections would render the standpipes unnecessary. They also argued that the standpipes were difficult to maintain, as they did not charge for their use, and they had no revenue set aside for this purpose. However, some of the women in al-Salamuniya did not accept these arguments and in 1993 they persuaded the Village Council to reopen one standpipe. As one woman put it: "We used to have several public standpipes in the village. Why were they shut down? They should be reopened for us to use. In our village we are accustomed to the use of the public standpipes."

In addition to piped water supplies, women in both communities could also use water from shallow handpumps. These were mostly situated on the canal bank so that surplus water could drain directly into the canal. The canals that were only a few steps away from most people's front doors were also an alternative source of water for some domestic activities. Although fewer than 3% of women questioned in our 1991–92 baseline census admitted to using water from the canals for drinking and cooking, many women (even if they had a water connection in the house) washed dishes and did their laundry in the canals, as we shall see in the next chapter.

Water quality—domestic supplies

Our water quality studies showed that there were marked differences in the chemical quality of water available for domestic use from piped supplies, handpumps and canals, as shown by a chemical analysis of water hardness, in table 8.2. Hardness is a characteristic given to water by salts, such as bicarbonates, sulfates and chlorides of calcium and magnesium. As all clothes washers know, it is much more difficult to dissolve soap in hard water than it is in "soft" water containing fewer of these salts. Hard water when used for drinking or for making tea tastes flat. As we shall see in the next chapter, women were well aware of these differences and took them into account when choosing sources of water for their domestic tasks.

Table 8.2: Chemical and biological quality of water from different sources in al-Garda and al-Salamuniya

Village	al-Garda			al-Salamuniya		
	Canal water	Main water tank	Hand pump	Canal water	Main water tank	Hand pump
Dissolved solids (mg/l)	200–260	580	510–770	200–430	760	790–910
Hardness (mg/l)	210–280	450	360–420	230–300	380	430–520
Faecal coliform (cfu/100 ml)	10^2–10^5	0	0–36	10^2–10^6	0	0–91
Plate count (cfu/ml)	10^2–10^9	300	65–3200	10^2–10^6	11×10^4	620–10^5

In the two settlements, the level of hardness was consistently lower for canal water than for piped water measured in the main holding tanks (and thus in tap water delivered to houses) and for the handpumps.

The level of total hardness of water from all sources in al-Salamuniya was slightly higher than in al-Garda. In both villages, the quality of the piped water and that from handpumps tested was within the legal chemical limits for drinking water; according to the Ministry of Public Health Decree of 1975, the upper limit for dissolved solids was 1500 mg/l and for total hardness 500.

The domestic water systems in al-Garda and al-Salamuniya are not unusual in providing hard water to their residents. These two systems rely, like most small systems in the Delta, on subsurface water. Engineers prefer to draw water from subsurface sources (even though they are highly mineralized) as they require only chlorination (a relatively inexpensive process) before distribution. Levels of hardness in al-Garda and al-Salamuniya were, in fact, lower than in many other villages within the governorate of Munufiya (K-Consult 1991: 24–33; El Katsha et al. 1989: 30). Moreover, subsurface water in southern Delta governorates such as Munufiya is, in general, less highly mineralized than that in the northern Delta. These regional and local differences need to be taken into account when assessing local women's decisions to use, or not to use, the waters of their local canal.

Bacteriological contamination of water is indicated by the presence of fecal coliform. Although coliform bacteria, even *E. coli*, are not necessarily of human origin, health authorities generally accept that their presence is an indication of unsafe water supplies, treatment failure, or the presence of excessive nutrients. If fecal coliform in a particular water sample *is* of human origin, however, it would indicate the possibility that the water might contain schistosome eggs. As indicated in table 8.2, although fecal coliform was absent in the main tanks from which the domestic waters supplies of our study villages was drawn, it *was* present in handpump water as well as canal water.

Toilets and latrines

Our baseline census indicated that the vast majority of households in al-Garda and al-Salamuniya (98.4% and 93.7% respectively), had their own toilets or latrines. However, the situation is rather different if one looks at the type of facilities in each household. It also looks rather different when one remembers that 100 households in al-Salamuniya (6.3% of the total) had no facilities at all.

Table 8.3: Type of toilet/latrine in al-Garda and al-Salamuniya 1991–92

	al-Garda		al-Salamuniya	
No toilet/latrine	19	(1.6%)	100	(6.3%)
Elevated w/flush	54	(4.6%)	95	(5.9%)
Squat w/flush	20	(1.7%)	22	(1.4%)
Squat w/out flush	850	(72.3%)	1267	(78.3%)
Hole	190	(16.2%)	106	(6.5%)
Other	42	(3.7%)	28	(1.8%)
Total households	1175	(100%)	1618	(100%)

Source: baseline census of households 1991–92.

The most common type of latrine in both villages was a squatting plate without a flush. These latrines were frequently near a tap or a container of water for anal cleansing, hand washing, and for pouring into the latrine. Some of these so-called "pour flush" types in al-Garda were actually linked to the sewerage system, but the owners had not yet found time to install an automatic flush, or did not think it was worth the expense. Only one in twenty households had a modern type elevated flush toilet. In al-Garda these were usually connected to the sewerage system. Others were connected to a septic tank that, hopefully, had sufficient capacity to absorb the large amount of water being flushed into the tank.

Contamination behavior

Schistosomiasis researchers have long held that the presence of latrines discourages people from defecating and urinating in or near canals, and hence transmitting schistosomiasis. This conventional wisdom needs to be tested by conducting studies of actual latrine *use*. In practice this is not easily done, given that most people consider their defecation procedures to be strictly private affairs.

It goes without saying that only if latrines (as opposed to canals and the "great outdoors") are used and hygienically maintained will they achieve their potential benefits for the health of the community. Given that *S. mansoni* eggs in the feces of just one individual, should they enter a canal, can perpetuate the disease cycle if they survive long enough to reach a vector snail, we found it worthwhile to ask local residents about their excretory practices. We believe that our discussions with village people on these sensitive issues respected their privacy, while at the same time yielding meaningful information (see Brieger 1994; Zumstein 1983).

In both villages, women and men said that very few people used the fields and canal banks for excretion. They pointed out that, because of increasingly intensive irrigated farming, the lack of protective vegetation, and new houses built in the fields, there were very few areas in the fields or near the canals that provided the necessary privacy for defecation in daylight hours. We certainly found little evidence of indiscriminate defecation along canals or in the fields. Thus the situation had changed markedly from the 1930s, when epidemiologists studying Delta villages, in which few households had latrines, noted evidence of indiscriminate defecation in the fields and on canal banks (Watts and El Katsha 1995; Farley 1991: 192–200).

Most women in our study villages who lived in households without latrines said that they used the animal shed for defecation, and many of the men said that they regularly used the bathrooms in the mosques. Few households used facilities at their neighbors' homes. Adults rarely mentioned that they used the fields, the canal, the canal bank, or an area near the house. However, in later discussions almost half of adults from both villages admitted that the children sometimes used the fields.

Water quality tests at sites where women washed laundry indicate that some contamination of canals with *E. coli* had occurred there. This was most likely when women washed babies and children's clothes soiled with feces. However, this was a relatively insignificant source of schistosomiasis transmission, because of the very low infection rate among children under five years old in both villages. Of the 205 children under five tested in 1992, only two children (both boys) were recorded as being infected (see chapter 7). This brings us to a consideration of how human waste was disposed of in our study communities; we begin by taking a closer look at the al-Garda treatment plant.

The sewerage system in al-Garda

The al-Garda treatment plant was provided by a grant from the government Organization for Rural Development (ORDEV). Built in 1988, its main objective was to lower the groundwater table in and around the village. It provided for the removal of domestic wastewater as well as sewage. One third of all households had connections to the sewerage system at the time of our 1991–92 baseline census. Given that a household had to have a water connection in order to have a sewerage link, this meant that, at that time, fewer than half of the households that could have been connected were actually connected.

Local records indicated that a large number of houses, 259, had been newly connected in 1990, and a further 90 in 1991. The sharp fall in new connections in that year was due to the high cost of installation. In our 1991–92 census, one quarter of the villagers said that they did not have a connection because of the high cost, and a further 38% said that they were too far away from the main pipeline and they could not afford to pay for their own access pipe. The cost of connections at that time ranged from under 100 LE (Egyptian pounds) to over 500 LE; 53 of the households connected (13%) had paid over 500 LE, a considerable sum for a rural household. In spite of these expenses, local records indicated that by the end of 1994 two thirds of the households were connected (i.e. 84% of those that had been linked to the water system three years earlier).

In al-Garda some of the residents cooperated with each other to provide sewerage connections for their households when the main sewer pipe was some distance from their houses. The men first contacted the village authorities and their neighbors who might be willing to share expenses. In one *hara*, a narrow alley in the center of the village, each household contributed 105 LE for a line running down the alley. One resident reported that his initial expense was LE 150, to construct a pipeline from his existing closed tank to the access line. Once the connection was made, he increased the number of taps in his house from two to six, and added an extra shower to the one he already had.

The Aqualife sewage and wastewater system installed in al-Garda was comparatively sophisticated, relying on an electric pump to provide a continuous flow of water to maintain bacterial action in the treatment pond. The system discharged the treated effluent into a drain half a kilometer from the settled area. However, our water quality tests at the discharge point showed that the treatment plant was not working satisfactorily, as the tests recorded a range of 3–4.5% removal of the biochemical oxygen demand (BOD). Translated into plain English, this meant that the bacteriological quality of the "treated" effluent was similar to raw sewage, and well above the limits for effluent set out in Law 48 of 1982. Treatment was complicated by the fact that the sullage (domestic wastewater) entering the system was very concentrated. This occurred because most women continued to use only small quantities of water for domestic tasks, and often reused it. In doing so, they were continuing the pattern of water use followed earlier, when they had to remove all water used in the house manually (see chapter 9; also Gemmell et al. 1991). In this case, effective treatment of the combined sullage and sewage effluent may have been beyond the capacity of the system.

What impact, then, did the sewerage system have on al-Garda? In informal discussions villagers said that they did not consider that the presence of

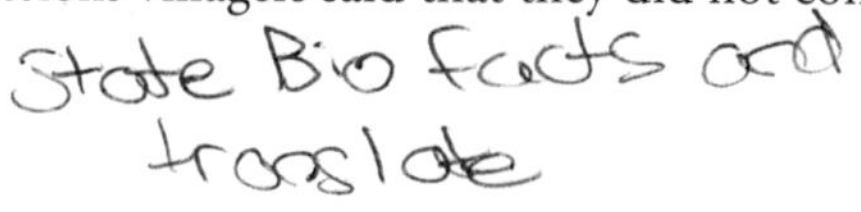

the system had influenced them in their use of the canals, which it to say, in engaging, or not engaging, in "risk" behavior. They all agreed, however, that it benefited *all residents* as it reduced the ground water level in the village (indeed that was its principal objective). This meant that now, compared to the recent past, there were fewer foul smelling waterlogged areas in low-lying places, fewer overflowing latrines and fewer damp houses. From the point of view of our research, we identified one benefit that was not specifically mentioned by villagers, namely that the sewerage system also reduced the number of latrines that needed to be emptied manually (and often very unhygienically).

Disposal of latrine effluent

Today, in crowded Nile delta villages such as al-Garda and al-Salamuniya, the careless disposal of latrine effluent may be a more important factor in the contamination of canals than indiscriminate defecation. At issue here is the number of latrines in both communities that require periodic emptying, and the facilities available to do this in a safe, hygienic fashion. Overall, one quarter of the latrines in al-Garda and 65% of those in al-Salamuniya needed to be emptied at some time or another, as shown in table 8.4.

Table 8.4: The evacuation of toilets/latrines in al-Garda and al-Salamuniya, 1991–92

	al-Garda		al-Salamuniya	
Evacuated	289	(25%)	985	(65%)
Never evacuated	473	(40%)	533	(35%)
Sewerage system	394	(35%)	0	
Total	1156		1518	

Source: baseline survey, households, 1991–92

The 38 latrines in al-Garda and 61 in al-Salamuniya that illegally emptied directly into a canal or drain obviously did not need to be emptied. In al-Salamuniya canals, when the water level was low the pipes were visible above the water line. According to some residents, a house site abutting a canal was favored because waste water and sewage could be piped directly into the canal. Other latrines in the two villages did not need to be evacuated because the holding tank was open at the base or porous; with the passage of the seasons, the level of the water in the soil moved up and down, thus removing the water and human waste deposited in the latrine. Neither, of course, was it necessary to empty the toilets or latrines in al-Garda that were linked to the sewerage system. But this still meant that two-thirds of the latrines in al-Salamuniya and one quarter of those in al-Garda required emptying. How was this done?

The official, approved way to empty a latrine was to arrange for a tanker truck to come from the mother village to vacuum pump the latrine contents directly into the large closed tank on the back of the truck. However, in practice, this system was of limited use. To begin with, many people could not afford it. Residents also reported that there were not enough trucks to serve all areas beyond the mother village. Moreover, trucks were too large to drive down the narrow alleys in the center of the settlement.

Accordingly, for the householder, the only realistic alternative was to hire a *soromati* from another village to empty the latrine holding tank. He lowered a bucket down with a rope into the tank, pulled it up and emptied it into a barrel on his donkey-drawn cart. It goes without saying that this was not a very hygienic procedure.

As of the early 1990s, there were no sites near either al-Garda or al-Salamuniya where *soromati* could legally dump the effluent they had collected. Dumping it into canals and drains was, of course, prohibited. The only permitted dumping sites were at urban sewage treatment plants at Shebbin al-Kom, the governorate capital, or at al-Marg, on the urban fringe of Cairo, over 40 km from al-Salamuniya—a long trot for a donkey drawn honey-wagon

Water quality—canals

The fecal contamination of canal water in our two study communities, combined with our evidence for the unsafe disposal of latrine effluent, suggests that these effluent disposal practices could introduce live schistosome eggs into canals and thus contribute to disease transmission. This possibility has not, until now, been discussed in the Egyptian setting. However, the infection of a group of school boys in Puerto Rico with *S. mansoni* after they had been swimming close to a sewage outfall suggested the possibility of disease transmission as a result of the improper treatment or disposal of sewage effluent (Clark et al. 1970).

In al-Garda and al-Salamuniya, schistosomes might survive close to pipes leading from latrines to the canals, or near a place where sewage has been dumped. Once in the water, the eggs change into early stage larvae, *miracidia*. They survive for about nine hours and during this time, moving briskly downstream, carried along by the flowing waters of the canal, they have to find a suitable snail host. In the snail host they undergo a form of asexual reproduction, producing many thousands of *cercariae* which are expelled into the water, often some distance from the point at which they originally entered the canal.

Between the spring of 1992 and the autumn of 1993, members of our team conducted quarterly biological and chemical tests of canal water within the built-up areas of al-Garda and al-Salamuniya and in the fields. We found that many of the canal sites where women customarily washed their dishes and did their laundry were contaminated with *E. coli*, as were sites close to pipes which discharged sewage into the canals, and where garbage was dumped. In al-Garda, some sites were identified where there was no obvious source of contamination. Contamination may have originated in subsurface water (polluted by unsafe latrines) seeping into the unlined canals, or it may have resulted from dumping by people living nearby. Several residents complained that people in nearby villages dumped sewage into canals that then flowed through their community.

Our water quality tests showed that in al-Garda, the four sites on the Kafr Tanbedi Canal, parallel to the main road, were less contaminated bacteriologically (i.e. with *E. coli*) and less chemically polluted than the five sites along the al-Garda secondary canal, which flowed through the center of the village. In al-Salamuniya canals overall pollution levels were higher. However, here there was a greater range in the contamination and pollution levels of the larger canals, the Diya al-Kom and the al-Salamuniya canals, that run through the center of the village.

Pollution levels varied considerably from place to place and over time, depending on when and where people threw sewage, waste water, and garbage into the canals, and when and where women working at the various sites added soap and other washing agents to the water. In both villages, some sites were so polluted, and had so little oxygen, that they could not support vector snails. In the three most highly contaminated sites in al-Salamuniya, where dissolved oxygen levels were almost zero, snail inspectors found only dead snails. In such cases, the high pollution levels acted as a deterrent to human water contact.

Snail vectors

Our surveys carried out in al-Garda and al-Salamuniya between April 1992 and May 1993 found vector snails of both *S. mansoni* and *S. haematobium*. What is most striking is the large number of *Biomphalaria alexandrina* snails (the vectors of *S. mansoni)*, at sites examined in al-Salamuniya, compared to al-Garda, as shown in figure 8.1.

Figure 8.1: Snail vectors in al-Salamuniya and al-Garda, 1993–94

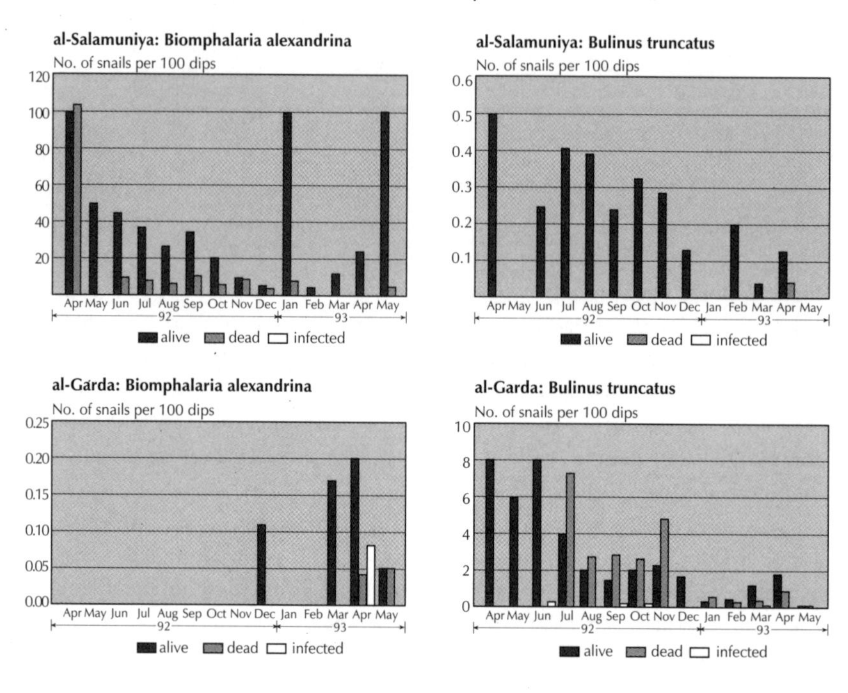

In al-Salamuniya, snail numbers were particularly high in April 1992 (233.3 live snails per 100 dips of the net and even more dead snails) and again in January 1993 (270 live snails) and in May 1993 (161 live snails). These three peaks occurred in early summer and just before the annual canal clearing in late January or early February. Our surveys found small numbers of snails in late February and March, and a modest number in August, which is usually the month when temperatures in the canal are highest.

Compared to the situation in al-Salamuniya, in al-Garda only very few *B. alexandrina* snails were found. They were not present at all during April to November 1992 or in January and February 1993. Peak numbers of snails were found in April 1993, a level of 1 snail per 500 dips. This contrasted markedly with the situation in the same month in al-Salamuniya, with its 233.5 snails per 100 dips (which works out at 1167.5 snails per 500 dips). Obviously, there is a close correlation between the smaller number of *S. mansoni* cases in al-Garda (compared to al-Salamuniya) and its much smaller host snail population.

In al-Salamuniya our surveys found only a few *B. truncatus* snails, the vector of *S. haematobium*, with a peak of 1 per 200 dips in April 1992; the only dead snails were found in April 1993. In al-Garda, we found more *B. truncatus* snails than in al-Salamuniya, with peaks of 8 live snails per 100 dips in April and June 1992. What was striking here was that in all the months from July to November 1992 the number of dead *B. truncatus* snails outnumbered those captured alive.

B. truncatus snails have generally been found to be less tolerant of biological and chemical pollution than *B. alexandrina* (the vectors of *S. mansoni*). Given the rapid rise of human population in Delta communities, and the concomitant increase in the chemical and biological pollutants reaching the canal, it seems very likely that *B. truncatus* snails are finding it increasingly difficult to survive. Indeed, in the mid-1980s, Cline and his team considered that the decline in the population of *B. truncatus* in the Delta was chiefly responsible for the decline in the prevalence of *S. haematobium* (Cline et al. 1989).

For both snail species, distribution was found to be related to the presence of vegetation. The snail population was higher where dense vegetation, especially *ward al-Nil*, water hyacinth, was found. Both villages had some stretches of canal that were almost completely covered by water hyacinth, making it very difficult for people to enter the water to wash dishes, swim or fish. Elsewhere, floating massing of water hyacinth moved slowly downstream on the current, with vector snails clinging to them, enjoying the nutrients and shade they provided. However, as dense vegetation affected the distribution of both snail types, it cannot be held responsible for any differences in density between the two species.

What is important to remember here is that snail populations fluctuate considerably over time and space partly because they inhabit moving water, but also because they have a short life span and a great reproductive potential (a feature which makes it difficult for mollusciciding programs to have a long term impact). We found considerable variability in snail density from site to site, as did Yousif in his study of *B. Alexandrina* snails in canals in Giza and Qalyubia governorates. He accounted for this variability in terms of the changing level of the water in the canals between irrigation rounds. The seasonal pattern of *B. alexandrina* that we noted in Munufiya governorate was similar to that noted by Yousif who also found fewer snails in canals just after the annual canal closure in January and February, and in high summer (Yousif et al. 1993).

On the basis of snail and water quality surveys, our team attempted to identify some canal-side activity sites in both villages that presented, on average, a relatively greater threat of infection to people than did other sites. Snail numbers were used as an indicator of the potential risk for people using such sites, as comparatively few infective snails were found. As a single infected snail can release thousands of *cercariae* into the water, it may only require one or two infective snails to endanger the people who visit the site.

Table 8.5: Likely and unlikely sites for transmission of S. *mansoni*, 1992–93

Site name	Fecal coliform /100ml	S. faecalis /100ml	Dissolved oxygen Mg/1	B.O.D. Mg/1	C.O.D. Mg/1	Live snails Bio	Infected snails	Activities
Likely Sites								
al-Khearsa	0	10^2–10^7	1–4	22–100	22–268	10–1000	Yes	Washing
Shaytani	0–10^6	10^3–10^5	0–3	14–70	38–100	0–100	No	
Mousely	0	10^4–10^8	0–8	16–60	58–100	4–335	No	
al-Diya al-Kom	0–10^3	10^3–10^4	0–7	12–80	58–116	0–1000	No	
Water tank	1–10^6	10^4–10^6	4–7	12–70	38–140	0–100	No	
Cemetery	0	10^4–10^5	0–3	20–60	58–140	0–3200	No	Washing
End Diya al-Kom Canal	10^2–10^6	10^3–10^5	3–10	20–180	40–288	1–200	Yes	
Unlikely Sites								
Sami's shop	0	10^2–10^5	5–7	19–90	38–150	0–33	No	Washing
Youth Center	0–10^2	10^2–10^3	5–8	12–70	29–98	0–7	No	
Beginning Salamuniya	10^3	10^2–10^5	4–10	20–100	40–211	0–33	No	
Health Unit	10^3–10^5	10^2–10^4	0–10	14–50	40–70	Zero	No	
Mina Fahmy	0–10^5	10^3–10^6	0–8	12–30	38–100	Zero	No	Washing

Snail and water quality information for sites in al-Salamuniya for 1992–93 indicated seven likely sites for transmission, and five unlikely sites, as shown in table 8.5. Two sites in the village which women used for washing were identified as likely sites of transmission. These were al-Kharsa, where some infected snails were found and the cemetery area. The other likely sites were in the fields *(zimam)*. Unlikely sites in al-Salamuniya in 1992–93 included two places the women used for washing: in front of Sami's barber shop (site 3 on the map) and in the *'izba* (hamlet) of Mina Fahmy, across the railway and the main canal from the built up area of the village. The site in front of the Youth Center, on the other side of the bridge from a women's washing site, was also an unlikely site. So too was the canal in front of the Rural Health Unit, in which no live snails were found, and where we recorded high rates of fecal coliform and very little oxygen. As indicated earlier, this stretch of the canal was so polluted that nobody would have wanted to go near it.

In the following year, 1993–94, the sites identified as being likely and unlikely to transmit infection were somewhat different. For example, these data suggested that the *'izba* of Mina Fahmy and the site in front of Sami's barber shop (identified as unlikely sites in 1992–93) were now among the most likely sites for infection. We found that the likely and unlikely sites differed not only for the two successive years, but also between the monthly surveys (and probably, indeed, on a day to day basis). This variation can be accounted for by variations in the amount of chemical and biological pollutants that find their way into the canal, and the flow of water that moves both snails and pollutants downstream.

For this reason, any posted notice warning that this or that site was especially unsafe would only be correct part of the time (and in any case some adult villagers would be unable to read such a notice). Therefore, it is especially important that snail inspectors are aware of local canal-side activity sites, test them as often as possible, and inform local people (and local health staff) immediately if they find infected snails (or even many uninfected snails) near such sites.

The irrigation system and schistosomiasis transmission

We can now rephrase our original question concerning the explanation for the difference in infection levels in the two communities, by asking why particular conditions favored the survival of large numbers of *B. alexandrina* snails in the canals in one community, al-Salamuniya, compared to the other, al-Garda.

From the 1930s onward, epidemiologists began to concern themselves with the possible impact on infection rates of seasonal basin irrigation, as compared to the more recently introduced perennial irrigation. They recognized that canal conditions under basin irrigation were less favorable to the survival of vector snails, partly because many snails died when the canals dried out, and the rapid flow of water when canals were full made it difficult for *miracidia* to find vector snails (Abdel-Wahab 1982: 63–64). We recalled these studies when we began to look at canal conditions in our study communities.

At the time of our study, we found that the canal regimes of al-Garda and al-Salamuniya were very different. In al-Garda, between irrigation rounds the main canal intakes were closed off and the canals were able to dry out, killing a high proportion of the snails. The usual alternating cycle, during which each farmer drew his water, was five days up and ten days down in winter and six days up and 12 down in summer. When the sluice gates opened the canals flowed rapidly.

In contrast, the canals in the system that irrigated the al-Salamuniya *zimam* flowed slowly during the time when the canal gates were open. Even when they were closed the canals never dried out, but always had some water and mud in the bottom. Unlike the canals in al-Garda, those in al-Salamuniya never dried out sufficiently to kill a large number of the snails. This system also resulted in some stretches of a canal becoming stagnant and highly polluted during periods of low water. This particular canal regime occurred because the whole canal system in al-Salamuniya was interconnected. Thus it was not possible to completely prevent the flow of water into or out of one stretch of canal, and the canals could not dry out between irrigation rounds or even during the annual winter canal cleaning.

Because of these differences, it appears that the canal regime in al-Salamuniya was more hospitable to *B. alexandrina* snails (the vectors of *S. mansoni*) than was the canal regime in al-Garda. On the other hand, we found that overall, canals in both communities had very few *B. truncatus* snails, the vectors of *S. haematobium*. The small number of *B. truncatus* snails found in al-Salamuniya (even fewer than in al-Garda) was probably due to the greater pollution levels of the al-Salamuniya canals.

Taking all the findings in this chapter together, it seems likely that local differences in the irrigation systems of al-Garda and al-Salamuniya can explain much of the difference in *S. mansoni* infections in the two villages. The impact of local differences in canal conditions on schistosomiasis infection levels is an issue that, so far as we know, has not been explored in other studies. However, to put this finding in perspective we need to look at the behavior of women, men, and children at the canals in the built-up areas of both villages and in the open fields. This is the subject of our next chapter.

9

Gender, Human Behavior, and Exposure to Schistosomiasis

Introduction

Epidemiological surveys of people who are infected with schistosomiasis categorize them according to age, sex, occupation, place of residence, and similar characteristics to provide clues about water related behavior that might be responsible for their infection. Some surveys go further and ask informants if they come into contact with potentially unsafe water sources, such as canals. There is a need to go beyond this, in order to investigate the ways in which people's canal-side behavior is related to what they perceive as functional necessity, and how it might also, at a deeper level, be due to their cultural and social attitudes. In short, we as social scientists want to understand *why* this "at risk" behavior occurs. Drawing on these findings we also seek to make some generalizations about the patterns of behavior associated with canal use and how these patterns can provide insights into how people acquire the disease in the local setting.

The epidemiological survey we undertook in mid-1992 in the two Delta communities, al-Garda and al-Salamuniya, showed that many more males than females were infected (much as has been the case in other Egyptian communities). Infection rates also varied according to age, with slightly different patterns for males and females, as discussed at the end of chapter 7. Addressing the question of why these differences existed and why prevalence rates were so different in the two communities, in the present chapter we will identify activities that took place within the built-up area of each village and in the fields surrounding the village. Then, drawing as much as possible on

their own perceptions, we will attempt to understand why local people persisted in using the canals and why, for that reason, schistosomiasis continued to feature in their lives.

Gendered space within the built-up areas

In the early 1990s, observant travelers passing through Egyptian village communities could see women, men, and children engaged in a variety of activities along the banks of the canals or even in the canal itself. In some locations, small groups of women were washing clothes or domestic utensils. At other sites, adolescent boys were swimming or playing around in the water.

Canal-side activities in built-up areas were segregated according to gender, with males and females carrying out specific types of activities in specific places. Observations at 13 canal sites within the built-up areas of al-Garda and al-Salamuniya in May and June 1992 showed that, during the morning hours at least, they were dominated by women doing their dishes or laundry, or washing grains or mats, as shown in figure 9.1.

Figure 9.1: Activities at village canals in al-Salamuniya and al-Garda

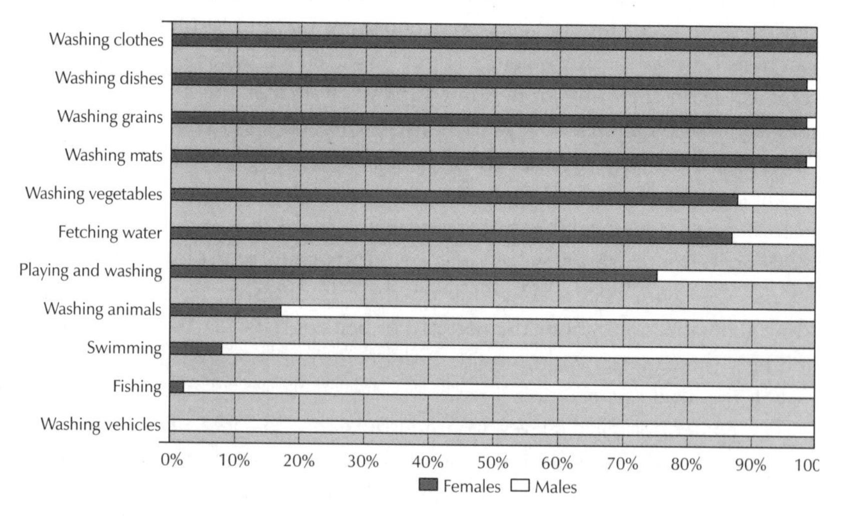

Source: observations at 13 canal sites, May–June 1992.

For these specific purposes women had identified some canal-side locations as more suitable than others. While doing laundry, in order to give their clothes a good scrub they needed stones on the edge of the canal and shallow water at its edge to stand in. They avoided sites with steep or unstable slopes as these were considered dangerous for the small children inevitably found playing near them. Three of these thirteen sites were also regarded by local women as convenient places for washing grain.

Women claimed use-rights to a particular site by gathering there to do their tasks every day. Whenever possible, they used canal sites in their own neighborhood, where they were already well known to dwellers in nearby houses. Regularly seen doing their washing, they felt no shame in exposing their bare feet, and even occasionally a flash of an uncovered arm or an ankle.

Women's perception of the space of their own lived world of the everyday, and of what lay beyond it, was also reflected in their head covering. Even indoors, most women wore a *mandil*, "handkerchief," a colorful cotton square wrapped around their head and tied in a knot on their forehead. When they went out of the house to visit a canal, they usually wore a *tarha*, a black net veil that covered the *mandil* and was looped loosely round the neck. Before leaving their own neighborhood, they put on an opaque black *hegab*, of cotton or synthetic fabric, to cover their head and shoulders.

When choosing a canal site for swimming, most boys selected a site with deep water and, if possible, a bridge or high bank from which they could dive. Accordingly, they seldom used the subsidiary canals that wended their way through the built-up center of their village. Men choosing a place to wash their trucks and cars also tended to use the larger canals. They parked their vehicles on a level area close to the canal, easily accessible to water for washing and rinsing.

In general terms, the canal-side sites within the built-up areas of a village were characterized by activities carried out either by groups of women and girls or by groups of men and boys. However, after the women had finished their morning tasks and gone home, the sites they had been using might be taken over by boys and men, and so cease to be regarded as women's space. Indeed, by late afternoon women tended to remain indoors and leave the spaces and paths near the canals to the locally resident men and boys.

During the morning hours, especially on market days, the streets and narrow alleys of the village core were filled with women and school children. The children were walking to school, and the women and their children who were not at school were leading water buffalo and donkeys to the fields. Later in the morning, men and women bought and sold produce at small shops and from locally resident street vendors. Especially on market days, peddlers (usually male) visited the village to sell clothes, shoes, household goods and agricultural tools. On these days, too, butchers slaughtered sheep, goats and cows and hung up whole carcasses on iron pegs outside their shops.

exactly where diff. pp. were at diff. times of day. Doing what

In al-Salamuniya, the principal road, which leads into the heart of the village, passes a school, crosses a major canal, and then leads into an open space bounded on one side by the youth center, as shown in figure 9.2. The youth center was the main gathering point for adolescent boys, especially during the summer holidays; the deep water of the nearby canal was attractive for swimming. Across the road from the youth center, women with large baskets of fresh vegetables, cheese and eggs gathered in lively groups every morning, looking for customers. On the other side of the canal from the vegetable sellers, other women were at work on the rough stone steps leading down to the canal, washing pots and pans and clothes. At the beginning and end of each shift at the school, children crowded the road, chatting and buying snacks from food vendors' carts.

Figure 9.2: Activity spaces along the road into al-Salamuniya

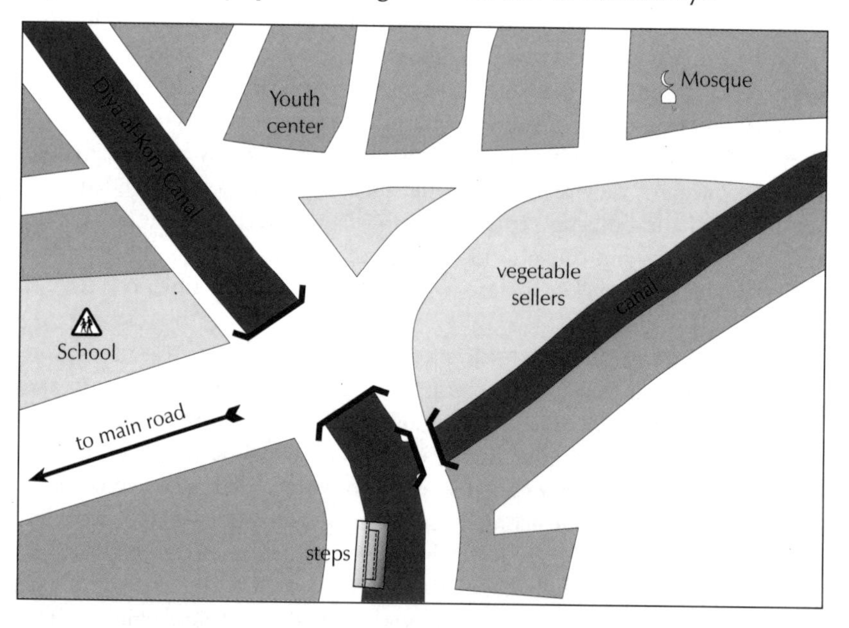

Gender, farming, and schistosomiasis infection

In view of the assumption that people engaged in farming are at risk of contracting schistosomiasis while irrigating their plots, we will begin by looking at the findings of our 1992 epidemiological survey. This found prevalence rates for full-time male farmers of 15 and over of 16% in al-Garda and 55% in al-Salamuniya. This was the highest rate for any occupational group. In al-Salamuniya, when looking at five primary occupational groups for residents over 15, we found significant differences in prevalence rates for these occupations (in five categories, $p < 0.005$). In both communities, many full-time farmers were older men, survivals from an earlier period when the proportion of farmers was much higher than it was in the early 1990s. In al-Garda, three of the farmers identified as infected in 1992 were over 60.

Full-time farmers were continuing to be infected during the years of our study. In our second annual survey in al-Salamuniya, taken in 1993, we found four new infections (incidence cases) among full time male farmers. Of those four, two were aged 30–34, one was 50–54, and the fourth was over 70. These findings were somewhat unusual, as most epidemiological studies have found a decline in infection levels with age, indicative of the development of immunity due to increasing age and/or exposure to infection (Gryseels 1994).

As we recognized at the beginning of the project that not all those who farmed did so full-time, we included in our baseline census a question which enabled us to identify people who worked in the fields after returning home from their other jobs, in the afternoon or on Fridays. Government jobs stopped at 2:30 p.m. and most of the plots in the *zimam* were only a short walk from the built up area of the village. For many residents who still owned or were able to rent a small plot of land, it made sense to farm it themselves rather than employ laborers.

Our 1991–92 baseline census indicated that although only a quarter of the males over 15 years of age described themselves as full-time farmers, another quarter said that they worked part-time in the fields every day. Similarly, a third of the women in the two communities indicated that they went to the fields for part of every day, although less than 1% (14 in each village) described themselves as full-time farmers (see table 7.5). In both al-Garda and al-Salamuniya, a third of the girls aged 5 to 14, and a slightly larger proportion of the boys also said they worked in the fields for part of every day. Given that our baseline census was carried out in December and January, relatively slack times in the agricultural year, these figures do suggest a high level of participation in farming in both communities.

Prevalence rates were higher for men and women in al-Garda and al-Salamuniya who reported that they worked full-time or part-time in the fields, compared to those who never did. In al-Salamuniya, of the 28 new cases recorded in 1993, 13 (46%) worked in the fields every day, compared to 8 (24%) who said they never did.

These figures suggest that full-time farmers and others who worked in the fields part of the time were at risk of schistosomiasis. Thus, our findings apparently differ substantially from those of the 1992 EPI 1, 2, 3 researchers who, in their published papers, failed to mention any specific findings about the relationship between farming and infection status. Their baseline questionnaire asked about the principal occupation, but did not inquire about part-time farming. Their neglect of these activities reflected a failure to recognize the extent of the occupational changes, associated with the increase in population and in non-farm employment, that had occurred in the previous thirty years in rural Egypt. Had they included part-time farmers, EPI 1, 2, 3 researchers may have been able to identify a statistically significant relationship between farming and infection, thus identifying farming as a "risk factor" for schistosomiasis. By the same token, by failing to ask about part-time occupations, they had no opportunity to assess the possible risks for women and children who worked part time in the fields, alongside men.

What farmers do—irrigation related behavior

Our observations along canals in the fields surrounding the two villages, during May and June 1992, focused on irrigation-related activities (rather than on planting, weeding, or harvesting). The main crops being grown were cotton and maize, with some vegetables in al-Garda. During these months, cotton had the heaviest water demands, requiring water at least every fifteen days. Observations were recorded wherever small family groups were working in the fields. Because these working areas were scattered and less regularly used than those in the village, it was not possible to identify specific sites for observation ahead of time. The difficulty of identifying regularly used sites in the fields (comparable to those in the built-up area) has doubtless deterred other researchers from carrying out observation studies of farmers. In the mid-1960s Farooq had apparently studied farmers' water contact in the fields but, unlike his studies in the built-up areas of the village, these were not, so far as we know, published (Farooq and Mallah 1966).

All of the observation sites were along the field canals, *misquas*. As these flowed below the level of the fields, during their irrigation turns the farmers had to lift the water to the field ditches that provided the water directly to their fields.

Farmers worked at irrigation sites from one to four hours or more, during the brief period when the water was available for their field. The median duration of a work session in al-Garda was one and a half hours, and in al-Salamuniya two and a half hours. Fifty-two observation sessions were held in each of the two village *zimam*. These recorded the activities of 348 people, of whom 40 (11.5%) were women. Observations were classified according to how much of a person's body was actually in contact with the canal water.

Table 9.1: Observations of exposure during irrigation related activities

	al-Garda		al-Salamuniya		Total	
	M	F	M	F		
Only hands	9.7%	63%	13%	42%	56	16%
Hands and feet	56%	37%	44%	38%	164	47%
To knee	34%	—	41%	14%	122	35%
Above knee	—	—	2%	7%	6	2%
Total	133	19	175	21	348	

Source: field observations May–June 1992: 52 sites in each village, N = 348

As shown in table 9.1, two thirds of the workers engaged in some phase of the irrigation process immersed no more than their hands and/or feet in the water, with most of the others only immersing themselves up to the knees, as shown in table 9.1. In both village *zimam*, during the observation period women were more likely than men to get only their hands wet. Much of the work done by women in al-Garda was done on dry land. Here, nine of the nineteen women were responsible for guiding a blindfolded buffalo or donkey operating a water wheel, *saqia.* In al-Salamuniya, where women had higher levels of exposure than in al-Garda, more than two thirds of the women observed (13 of 21) immersed their hands *and* feet in the water.

In al-Salamuniya, portable diesel pumps were in more common use than in al-Garda. Forty (75%) of the 52 observation sessions there involved these pumps, compared to only 23 (44%) out of 52 sessions in al-Garda. Although a portable diesel engine is more user-friendly and "modern" than a fixed water wheel, it compels farmers to spend more time in the water than if they were using a water wheel; it was in al-Salamuniya that 55% of full-time male farmers were infected. In order to keep the canal water flowing while using a diesel pump, the operator had to stand in the feeder canal, firstly to install the pump, and then to lift out the floating garbage and weeds in the feeder canal that would otherwise block the hose. These were very time consuming tasks.

In addition to the actual processes of irrigation, part-time and full-time farmers engaged in other activities that brought them into contact with canal water. Observers noted that many washed themselves clean of dirt and sweat after completing their farming tasks. Thirty-six percent of the observed male workers and 48% of the women washed up in the canal. Health specialists know that the penetration of canal borne *cercariae* is much reduced if people towel themselves immediately after contact with infected water. However, only 22% of the men observed washing in the canal and 28% of the women used a towel immediately after washing. Washing up under a handpump would be risk free, as far as *cercariae* were concerned. In our observation sessions, 37% of the women (15 out of 40) and only 18% of men (56 out of 318) used a pump for washing up.

Farmers' attitudes toward water contact

Focus group discussions with farmers (conducted after we had finished all our observation sessions) revealed that all of the male full-time farmers were aware of the risk of exposure to schistosomiasis. Yet they argued forcefully that there was no way they could avoid contact with canal water, particularly if they used diesel pumps. The old ways were not seen as "risk free" either. Mahmoud, an elderly farmer, said: "If we use lifting pumps or traditional methods to irrigate we have to go into the canal, either to clear the way for the pump or to change the course of the water."

Farmers who used diesel pumps to irrigate cotton fields said that during summer they often needed to spend four to five hours a day in continual contact with the water to do their necessary tasks. They complained that there was more rubbish, especially plastic bags, in the canals than there was when they were young and that as a result they had to spend more time in the water preventing the irrigation hose from getting clogged up. In al-Salamuniya farmers also complained that there was less water in the canals, and a slower rate of flow than in the past, so that they were compelled to spend longer standing in the feeder canals controlling the flow of water to the small field canals.

Most farmers knew that they risked schistosomiasis when irrigating their fields. At least one blamed people in the village for putting disease-causing agents into the water. As Kamel, an elderly farmer, stated: "Going down to the canal is a must. However, this is not the problem. The problem is that the Nile water comes in clean, but once it passes through the village it becomes polluted by the people. If the canals are only used for irrigation there will be no problem of schistosomiasis." Some elderly farmers took a different view of the risks of schistosomiasis and claimed that they had never been infected. As Saad stated: "All my life I have been irrigating and I have never had schistosomiasis. God is the protector."

The farmers we talked to in our focus group discussions could not think of any practical way in which they could protect themselves from exposure to schistosomiasis. We asked them about the possibility of wearing high plastic or rubber boots. In response they said that although they knew very well that such boots would block the passage of the disease causal agents (that we know as *cercariae*), boots were hot and uncomfortable in summer, and tended to become stuck in the mud. The farmers we talked to were also skeptical of the value of a barrier cream, which had to be spread on all exposed parts of the body before water contact. In separate discussions, women who farmed told us that they also accepted the risks of exposure, and saw them as unavoidable (Watts and El Katsha 1997). In chapter 13, we will discuss alternative approaches to the problem of farming exposure that involve interventions to change the ecology and regime of the canals and control the intermediate snail vectors (rather than focusing on changing farmers' behavior).

Women's domestic tasks at the canal

Observations at water contact sites

For most women in al-Garda and al-Salamuniya, making use of a nearby canal when doing domestic tasks was as much a part of their life as working in the fields. Our observations of women's canal-side activities provided considerable information about the nature and frequency of these activities and how much skin they exposed to the water. Later follow-up discussions helped us to understand why they persisted in using the canals.

The six canal-side sites where we carried out our observations in al-Garda and the seven in al-Salamuniya all had local names, indicating that they were well known and commonly used. Their locations are shown on figures 9.3 and 9.4. In al-Garda, the names of all sites incorporated the word *murada*, steps. One of the more popular sites, where wide concrete steps lead down to the canal which flowed through the center of the village, was called Murada Abu Shabaka. It was adjacent to one of al-Garda's few remaining functioning handpumps, and had long served people living nearby as a center of sociability.

During the morning hours, at any one time during the period of our observations, from two to nine women were working at each canal site. The composition of the group changed frequently as women, with young children in tow, came and went. The most common activities we observed (and that women reported in the baseline census) were washing dishes and washing clothes, as shown in table 9.2. These two tasks were almost exclusively the domain of women and girls.

In order to make full use of its water, women had to stand in the canal part of the time. However, the amount of skin they exposed to canal water varied according to the activity involved. Our analysis of the observations found that on nine out of ten occasions, women who were washing dishes only put their feet and hands in the water. A similar level of exposure was recorded for most of the women who were doing the family laundry, the second most frequently observed activity. In these instances, women using canal water were not exposing a great deal of skin, but they were doing so relatively often. In contrast, as we shall see, women who were washing grain necessarily had to stand waist-high in the water, but they did so less often.

Figure 9.3: Water contact sites at al-Garda

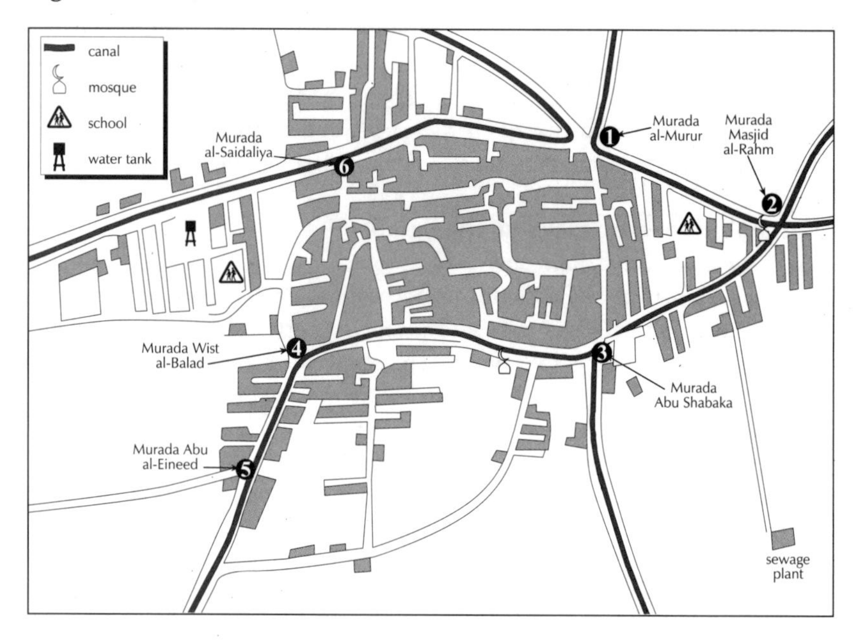

Figure 9.4: Water contact sites at al-Salamuniya

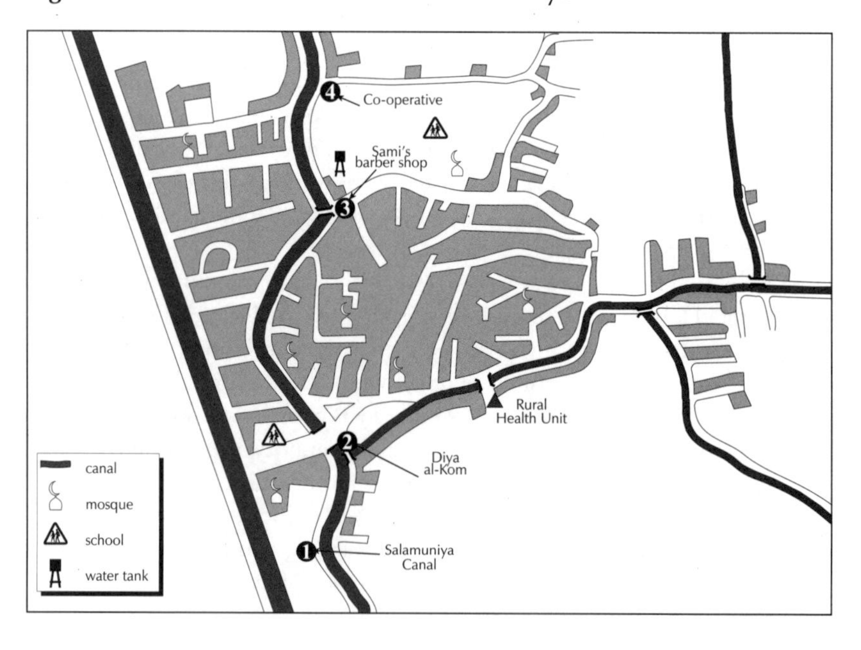

Table 9.2: Activities observed at village canals, 1992

Activity	#	%	% female	% <15	hands and feet	bare feet
Predominantly female activities						
Washing clothes	1641	18%	100%	25%	90%	21%
Washing dishes	3525	39%	>99%	42%	93%	18%
Washing grain	681	7.5%	>99%	8%	21%	32%
Washing mats	780	8.5%	>99%	27%	83%	13%
Washing vegetables	26	<1%	88%	38%	100%	4%
Fetching water	783	8.5%	87%	17%	99%	16%
Playing, washing	513	5.5%	75%	52%	85%	25%
Predominantly male activities						
Washing animals	98	1%	17%	7%	79%	50%
Swimming	781	8.5%	8%	81%	0%	95%
Fishing	142	1.5%	2%	79%	75%	49%
Washing vehicles	50	<1%	0	28%	100%	47%
		(100%)				

Number of activities: 9020 at 364 sessions; see Watts and El Katsha 1996.

Washing dishes was the most frequently observed activity at the canals. The task was usually done in the late morning or early afternoon, between 11 and 2 o'clock. Most of those observed washing dishes were young unmarried girls, and 14% of them were under ten years old; within this age group, this was the most frequently observed activity. However, not all mothers were satisfied with their daughters' performance. Soheir, for example, said she could not trust her eight-year-old daughter not to drop the breakables or to allow the light plastic dishes to float out beyond her reach. She gave these as the reasons why she did not allow her daughter to help with the canal-side washing. Soheir did not mention that by staying away, her daughter avoided the risk of contact with schistosomiasis-infected water.

Women from large families might spend up to two hours a day washing dishes in a neighboring canal. The busiest dish-washing day was usually Friday, when dishes from the main family meal of the week, eaten in the late afternoon or early evening on Thursday, at the end of the work week, had to be washed.

After a meal, the women and girls piled up the dirty dishes in a large aluminum or enamel basin (known as a *tisht*) in the hall of the house (near the entrance door), ready to be taken to the canal. At the canal, after giving the dishes a preliminary rinse, they washed them with soap and water. They washed the glassware and cutlery first, with soap and rice hay. Food encrusted plates and cooking pots were rubbed clean with a loofah (Arabic *lifa*), or with a pad made of clean straw or fibers taken from the trunk of a palm tree. They used mud or ashes to clean very dirty cooking pots. After thoroughly rinsing the soap off tableware, cutlery and pots until they became bright and shiny, they put them at the side of the canal to drain. Then the girls and women carried them home in large basins carefully balanced on their heads.

Washing clothes, during the period of observation in our study communities, was usually done in the morning, between 9 and 12 o'clock. In most households the laundry was done once or twice a week, or more frequently if the family was large or there were many young children. Older girls and young women (rather than children) did most of the laundry as it required lifting heavy loads. However, girls as young as ten years old helped their mothers at the canal side, learning how to do everything "just right."

Women brought their dirty laundry to the canals in aluminum or enamel basins. While washing, some of them squatted on the steps or on stones above the water, or on the sloping canal bank; others preferred to stand in the shallow water. They first soaked dirty clothes in a basin of soapy water, sometimes adding a stain remover. Then, they scrubbed the clothes on stones, after which they swished them vigorously around in the canal's slowly flowing water to rinse them. After wringing out the clothes by hand (so wetting their arms as well as their hands), they took them home. There, the wet clothes were hung out to dry on a line strung outside the window or on the roof. Sometimes clothes were spread out to dry on clean straw (see also El Katsha et al. 1989; El Katsha and White 1989).

According to our baseline census (1991–92), more than 80% of the households in the two study communities had a light, portable electric washing machine. These machines were not fully automatic, and consisted of a small, single chamber in which the clothes could be agitated (this replaced the process of beating the clothes against stones). Often the housewife soaked the clothes in a large basin before putting them into the machine. At the end of the cycle, she removed the clothes and rinsed them in a basin of water. As rinsing required a large amount of water, some women took the clothes from their machines directly to a canal to rinse them. Housewives who lived very near a canal sometimes used a long extension cord and brought their machines down to the edge of the water. In this way they were saved the trouble of carrying a load of heavy, wet clothes from their house to rinse in the canal.

Washing mats and other heavy articles such as blankets, quilts and coverlets was a particularly strenuous job. In our study communities, in 1992, it was mostly performed by the young, strong married women who were regarded as the "maids of all work." Forty-four percent of the women observed washing this heavy laundry were between 15 and 25, the highest percentage for any activity in this age group.

The washing of heavy items such as these was done only once or twice a year, usually on a fine spring day. Because it required a great deal of water, young women found it much easier to do it down at the canal, rather than in basins at home.

Washing grain in large quantities was also a seasonal activity, done mostly in late spring, after the wheat harvest. In former times, grain washing was a frequent activity in preparation for baking bread at home. These days, however, village people find it far more convenient to buy subsidized bread rather than baking their own. (Nevertheless, in our study communities some women still baked bread as they were convinced that the bread they baked themselves was much tastier and of a better quality than the subsidized bread. Members of the research team can vouch for its superior quality, as they have shared newly baked bread and sharp soft cheese, *mish*, with village women.)

In the early 1990s, grain washing was usually done by two women working together. After they brought the grain to the embankment in a large basket, one woman remained on the canal bank and handed a small basket of grain to her partner, who had entered the water fully clothed. The woman in the canal then washed the grain by vigorously shaking the basket in the water so that the light, defective grains, as well as the dust and pesticide residues that had accumulated during storage, rose to the surface and floated away. This required skillful maneuvering, as the top of the basket had to be kept just at water level so that all its contents did not float away. When the two women had finished washing their grain, they carried it back to the house where they spread it out in the sun to dry, either on the roof or on mats in the courtyard.

Grain washing was an occupation considered appropriate for women over about 35. Young girls were not allowed to wash grain in case they let the basket, and the grain, sink too far below the surface. To drop the basket while it was in the water would, of course, mean that all the contents would be lost.

Among the women standing in the canal up to their waist for an hour or more washing grain, there was considerable risk of acquiring schistosomiasis. Their modest, but billowing cotton galabiyas (and whatever else they wore underneath) would be unlikely to prevent some *cercariae* from penetrating their skin.

Fetching water was a fairly frequent activity during our 1992 observation sessions. Women fetched small quantities of water from the canal to use at home. Most women and girls only briefly wetted their hands, and maybe their feet, while leaning into the canal to fill their water pot.

Washing vegetables in canals in the built-up areas of our study villages was a relatively minor water contact activity. Vegetables that were to be spruced up for marketing were frequently dipped in a basin of canal water; this could be done in a few short minutes. In any case, most women we talked to were aware that it was more hygienic to wash off vegetables that were to be eaten raw in tap water rather than in canal water.

Why do women continue to use the canals?

In the early 1990s, our focus group and household discussions revealed that most women in our study communities regarded the use of canals as unavoidable, given local conditions. It was a form of behavior everyone took for granted. As one woman with a secondary school education said: "We are like fish. If we leave the canal we will die." Nabila, another educated woman who had a water connection in her house said: "Yes, my daughter had schistosomiasis and she took the treatment. But it was no use. She went back to the canal. We cannot do without it."

Only a few of the women we talked to suggested that their educated husband thought it was not a good idea for her to do laundry or dishes in the canal; they did not say whether this was because the husband thought it was archaic "country" behavior, or because it was unhealthy. Other women admitted that, since they were in full-time employment themselves and were very busy, they paid another woman to do the dishes, knowing that she took them to the canal.

In addition to the general consensus that it was standard behavior to use the canal for washing dishes and clothes, women in al-Garda and al-Salamuniya provided more specific explanations for their behavior. They did not, however, see them as a set of discrete "reasons," but rather as post facto rationales for what was locally "taken for granted" and customary.

Saving effort

Local women said that washing at the canal saved them time and effort, even if they had water connections in their house. It was far easier to do dishes and to scrub and wash clothes in the canal than in their kitchen. Washing at home required many basins, all of which had to be filled and emptied by hand. In contrast, while washing in the canal they would only need one or two basins to carry the clothes and dishes to and from the canal. Then too, in terms of the quantities of water required, it was much more convenient to wash heavy things like blankets and quilts directly in the canal than at home.

To an outside observer, washing at the canal looks like hard work. However, our women informants said that it was certainly no more strenuous than washing at home. In both locations, it was customary for women to squat while doing their work. Few houses had a raised sink and draining board, or a water tap located above a drain. Women whose kitchen had only a single tap and had no drainage outlet had to fill all the washing basins they needed, and afterward throw the used water away outside the house, in the street or in a canal. Women without a tap in the house had to carry in all the water they needed for the family personal and domestic needs.

As was the case with dishes and laundry, washing grain in the canal was also perceived as saving time and energy. Women said that they found it exhausting to do this at home, and that however thoroughly they washed the grain in basins on the kitchen floor it never seemed to get really clean. They claimed that it took them two days' work at home to wash the same amount of grain they could wash in two hours at the canal.

Access to drainage or sewerage systems

Women in al-Garda living in houses that were attached to the sewerage system could easily dispose of their waste domestic water by pouring it down the drain in their kitchen or down the toilet or latrine. Women in either of the study communities living in a house that boasted a "safe" septic tank could pour some of their waste water there, but if too much sullage was poured in, the septic tank might overflow. Most other types of effluent containers in use in the two study communities were also likely to flood if women poured in sullage. Thus, women living in houses that were not connected to the sewerage system, or did not have an efficient septic tank, found it easiest to throw their wastewater out of the door or to carry it to the nearest canal.

For women living in mud-brick houses of the sort common in the heart of the built-up areas of our two study communities (58% of households in al-Salamuniya and 25% in al-Garda), washing at home was especially hazardous. Women complained that when they washed clothes in a basin on the floor, the water splashing around wetted the floor and the lower part of the walls. Over time, this water seriously weakened the mud bricks, and ultimately could lead to the collapse of the building. Thus from a practical point of view it was entirely logical for women living in mud-brick houses to wash their laundry, their grain, and their dishes in the canal, rather than on the kitchen floor.

Our study appears to be the first to link the potential damage to a mud brick house with women's use of the canal for washing. Researchers may have neglected this relationship because they did not ask the women themselves why they use canals. Farooq and his colleagues considered mud brick houses of importance primarily as an indicator of low socio-economic status in the poor, rural communities he studied (Farooq et al. 1996 B).

Sociability

Perhaps most important of all, we found that canal-side washing places were centers of sociability for women in our study communities who might not otherwise have had an opportunity to leave their homes. As Fatma, a young married woman who had a sewerage connection in her house, said: "Washing dishes in the canal is entertaining, because one is not alone; many people are around. Also the light of God is better outside." In short, these canal-side meeting places provided legitimate opportunities for women (away from the prying ears of men) to discuss intimate topics, and to catch up on local gossip.

For adolescent girls and young married women, visiting the canal provided one of the few socially sanctioned occasions when they could show off their domestic prowess. A young woman could attract the attention of a potential marriage partner or a future mother-in-law as she worked assiduously at the canal and on her way home walked proudly through the village with a pile of gleaming dishes or clean clothes balanced on her head. Certainly, mothers spent a lot of time discussing possible marriage partners for their children, for they viewed marriage as a link between families as much as between individuals. Thus, it is not surprising that mothers often encouraged their daughters to wash at the canal, despite health warnings on TV and radio. If girls wanted to go to the canal with their friends, most mothers did little to restrain them.

Water quality

In contrast to the water they obtained from a standpipe or a tap in their own house, community women described canal water as *hilwa,* "sweet." Because it was soft, it made a lot of soapsuds (hard tap water did not) and got clothes whiter and aluminum dishes brighter than the tap water (in this respect women's perceptions of canal and tap water were in accord with the chemical analysis discussed in the previous chapter). Part of a housewife's pride was to see the men of her household in sparkling white *galabiyas*, and children and womenfolk in bright colored clothes.

Some women washed their hair in canal water, which they believed nourished it and made it grow and glisten; for modesty's sake, they carried the water from the canal and washed their hair indoors. Although women from al-Garda and al-Salamuniya rarely mentioned using canal water for cooking, they admitted that some food tasted better if made with water from that source. The most common example was that tea made with canal water tasted better than tea made with highly mineralized tap water. They said that *fuul*, a popular dish of slowly boiled fava beans, should also be cooked in canal water. They also preferred canal water for making pickles, eaten along with the main meal of the day.

A few women told interviewers that they did not regard the canal water as "clean." It seems, however, that they were thinking not of the *cercariae* searching for ways to enter their body but, instead, of the microbes (a word that has entered colloquial Arabic as *mikrubat*) that caused diarrhea and dysentery. They confessed that, after they had washed their dishes in the canal (with its risk of schistosomiasis) they brought them back to the house to rinse them in tap water.

Cost

Some women told interviewers that the high costs of metered water (coming through a tap) and, in al-Garda, the additional charge for the use of the sewerage system had caused them to go back to using the canal (which, as they said, was free). Their concerns were economically sound. At the time of our survey, the water pipes leading to houses in al-Garda and al-Salamuniya were in the process of being fitted with meters; householders then paid according to the amount of water they used, rather than the previous flat rate. Within a few months of the installation of the meters, the cost of water had jumped from 9 to 23 piasters a cubic meter. In al-Garda, households that were attached to the sewerage system found themselves burdened with an additional payment that amounted to 50% of their water bill. Everyone dreaded the time when the quarterly bills fell due. In al-Garda in 1994, women in large households with both water and sewerage said that the charges now amounted to 100 Egyptian pounds (approximately $30) a quarter, compared to 10 pounds a quarter in previous years. Finding money for prompt payment of this bill was difficult for rural households with an irregular cash flow.

Other activities at the canal

Playing and washing

The Arabic term *bistahama* refers to a wide range of water related activities. These include children playing and women washing themselves after their work. Although it was not considered appropriate for women and girls to swim in the canal, it was acceptable for them to go into the water fully clothed for a thorough wash. As they walked home their long loose dresses dried in the breeze, and they felt refreshed.

Among young people under ten years old, playing and washing in the canal accounted for the largest proportion of the observed activities of that age group, one third of the total in 1992. Children too young to go to school played, under the watchful eye of their mothers or older sisters, on the canal bank or more rarely, in the water. However, the fact that no girls and only two boys of the 205 children under five who were tested were infected, suggests that their actual water contact was very limited. When children reached school age, they were expected to play in gender segregated groups. The boys were allowed to play and swim in the canals without direct parental supervision, but girls were expected to be with their mothers at the canal side doing domestic tasks, learning their future domestic roles.

It is obligatory for Muslims to perform ritual ablutions, *wadu'*, before prayers; this involves washing the body orifices, face, hands and feet. In our study villages most men performed their ablutions in the bathrooms attached to the mosques. But a few men, who were working in the fields when the call to prayer came to them from the loudspeaker of a distant mosque, performed their ablutions using water from a nearby canal, before rolling out their prayer mat to face Mecca and kneeling down to pray.

Swimming in village canals

Researchers have often found a high rate of infection among school-age Egyptian boys in rural settings. In a study in Upper Egypt Helmut Kloos and his colleagues (1983, 1990) traced this directly to their fondness for swimming in canals, but we do not know how much time these boys spent farming. In al-Salamuniya and al-Garda we noticed many groups of boys swimming in the canals so we decided to supplement our earlier study at the thirteen village sites with a brief observation study in al-Garda and to follow these up with short discussions with the boys.

At the thirteen canal sites within the built up area of the two villages, we found that the great majority of those observed swimming were boys under 14; 20% were under ten years of age. They swam mainly at three of these sites. At the al-Garda traffic station, named Murada al-Murur, the boys liked to jump from the bridge into the canal and show off in front of passers-by on the main road. Not unexpectedly, swimmers had the highest proportion of the body immersed of all activities observed; most of the older boys were diving and swimming underwater.

Later in the summer of 1992, at four popular swimming sites in al-Garda, we observed groups of boys aged between about 6 and 15. They said that they swam every day during the long summer vacation, especially after working in the fields. As one young boy, Mohammed, said: "When we work in the fields we get itchy from the worms [from the cotton plants], so we swim."

The boys especially liked to swim during the days when the water was high, flowing faster, and cleaner than usual. They played ball games in the water, jumped from the bridges and the riverbank, and raced each other. The older boys wore underpants or swimming trunks, but the younger boys often swam naked. Many of the boys admitted that their parents did not approve of them swimming because of the danger of schistosomiasis and the danger of drowning during days when the water was deeper and flowed more rapidly than usual. But many fathers, remembering the youthful joys of swimming in the hot summer weather, agreed with their sons that there were few alternative opportunities available during the long summer vacation and even taught them to swim. Boys also told us that they swam in the larger canals flowing through the fields, often on their way home from farm work. Here they could play around without being seen by adults. A favorite place on the outskirts of al-Garda was a secluded spot near the al-Shat praying place.

Fishing

In the study villages, fishing was a part-time activity and a recreation, generally using a simple pole and line. Fishermen stood on the canal bank or waded in the shallow water near the edge of the canal. Many of those observed fishing in 1992 were boys; 94 of the 142 (66%) observed in the two communities were between 10 and 15 years old.

Washing animals

Few people in large villages such as al-Garda and al-Salamuniya kept large livestock. Accordingly, nearly half of the animal washing activities we observed in 1992 took place at the site in the hamlet near al-Salamuniya. Seventy-nine percent of all these activities only involved the owner immersing his hands and feet in the canal. However, men occasionally entered the water fully clothed to wash water buffalo.

Washing vehicles

In 1992, the practice of washing trucks, taxis, and minibuses near village canals, and even in a few inches of water, was not particularly common. Most of those who did so were adult men, rather than adolescents or young boys. The most frequented sites were on or near the major through road, near vehicle repair shops and the transport routes for taxis and minibuses.

Infection rates and canal use

Overall, the patterns of canal activities we observed in 1992 in al-Garda and al-Salamuniya were similar. The people we interviewed said that using the canals was accepted as an inevitable part of village life, passed down from generation to generation. We found that many residents, women, men, and children, could not see an alternative to using the canal in certain situations. They were used by both illiterate and more educated people. This helped us to understand why our epidemiological survey had failed to find a statistical correlation between educational level and infection (neither did the 1992 EPI 1, 2, 3 study find this correlation).

For women, the most important explanation for canal use was sociability. In addition to this, some women necessarily had to use the canal because they lacked adequate facilities at home. Many lived in houses that were not connected to piped water. Even more numerous were the women whose houses did not have a convenient or safe drain into which they could pour their sullage. For women who lived in mud-brick houses, this problem was particularly acute. Yet even for women living in houses made of fired brick or concrete, and with piped water and drainage, the canal fulfilled a deeper function. Here, they could legitimately meet their friends and compare notes about family life and family problems.

Our observations show that most of the domestic work at the canals is done by young women before they are married and in the early years of their married life. In both villages, the highest infection rates for women are found in the age group 15 to 34, suggesting that domestic exposure is, indeed, responsible for much of the infection among women in this age group. In al-Salamuniya this group accounts for just over half (54%) of all cases, with 37% infected (compared to an overall rate of 20% among women). In al-Garda, while the overall infection level is lower, the figures tell the same story. Sixty percent of those infected (17 out of 28) are between the ages of 15 and 34, an infection rate of 7.2% compared to 5.5% for all females.

In the case of school-age and adolescent boys, the canal was the obvious place to swim and play, especially during the hot summer months, often after a long morning's work in the fields. However, our prevalence survey (in mid-1992) found that boys aged 6 to 15 had a much *lower* rate of infection than did older males, even though some of these young people said they were swimming after working in the fields.

We have seen that farmers, regardless of age, had higher infection rates than other occupational groups. Among full-time and part-time farmers working in the fields, who necessarily had to adjust the flow of water as it moved from the *misqa*s, field canals, along the field ditches, contact with water was inevitable. This contact lasted longer if the operators used a modern diesel pump than if they used an old-fashioned water wheel.

These complex patterns of farming and domestic exposure to canal water, at different stages in the life cycle of females and males, are a major challenge for those seeking to reduce the impact of schistosomiasis upon the population of Egypt. As we have seen, since 1988, rural people who came down with schistosomiasis can be treated with a single dose of praziquantel and (in between six and eight cases out of ten) can be cured. Yet at present, most village people see no alternative but to continue to use the canal, even if this means that they risk becoming reinfected.

In this chapter, we have explored the first stage in the schistosomiasis infection process in human beings, exposure to infection, by looking at groups of people, activities, and places likely to be involved in this process. In the next chapter we will look at villagers' knowledge of schistosomiasis and their choice of sources for diagnosis and treatment.

10

Local People's Understanding of Schistosomiasis

Introduction

As understood by medical science, schistosomiasis is a disease caused by a pathogen that spends part of its life cycle in the human body. When the worm reaches the bladder or intestines (depending on whether it is *S. mansoni* or *S. haematobium)* the worm and eggs are associated with signs and specific symptoms that a physician can identify. So long as the worm is present in the body, it produces eggs which are excreted in feces (in the case of *S. mansoni*) or urine (in the case of *S. haematobium*); the identification of eggs in the urine or feces is the most common way to identify an infection. Laboratory analyses of blood samples can also identify current and past infection. In this chapter we will show that this biomedical view is somewhat different from the way the disease is understood by non-medically trained Egyptians.

When we started our research, we recognized that Egyptians had long been accustomed to using primary health care facilities for the diagnosis and treatment of schistosomiasis, and had some idea of what to expect when they got there. However, we wanted to know more about this local knowledge and what happened *before* a person arrived at the health facility. This short chapter provides a background for the next chapter, which explores what happened when villagers went to primary health centers (principally that in al-Salamuniya) for diagnosis and treatment.

Knowing about bilharzia

In all our discussions with local people in the two study communities, we rarely found even a child who had never heard of *harzi* or *bilharzia*. Of 296 adults interviewed in 80 households, only two had not heard of the disease. As Fatma said: "We grew up hearing about it." Residents did not consider there was any stigma attached to having the disease; it was something that happened to nearly everyone at one time or another in their life.

When we asked how one could get schistosomiasis, 95% of the women and 95% of the men mentioned the canals as a source of the infection.

Table 10.1: Villagers' knowledge of how one can get schistosomiasis

Do not know	5	
Mentioned the canal	287	95.5%
Only from the canal	129	44%
More than one cause	153	52%
Dirtiness and polluted water	6	
God's will	3	

N = 296, multiple responses
Source: household survey, 1993

However, as shown in table 10.1, slightly more than half (52%) of all respondents mentioned *more* than one way to become infected. When they did so they often used the local term *mikrubat* for any "microbes" that were invisible to the naked eye, but were known to cause a disease or diseases. In addition to exposure to canal water, respondents frequently stated that schistosomiasis could be picked up through the soles of the feet when going barefoot, or by eating unwashed vegetables. They probably based these perceptions on incorrect recollections of earlier official health education messages. The belief about going barefoot was probably confused with an earlier message about hookworm that encouraged people to wear shoes. Until the 1950s, campaigns against hookworm and schistosomiasis originated in the same section of the health department (Khalil 1949) and they could have been confused in the minds of some of the listeners.

Few of our informants in al-Garda and al-Salamuniya knew much about the complexities of the schistosomiasis transmission cycle. For them, the disease was simply in the water. As one woman explained: "Schistosomiasis lives in the canals." Some, however, did mention the existence of a worm, *duda*. Given that the worm existed only *inside* the human body, this idea would also seem to have come from a health message. Indeed, one of the health education spots on TV during our research was about *al-duda al-aziya*, "the troublemaker worm." Children playing in the canal would often sing this jingle when they saw us walking along a canal-side path. In so doing, they were both recognizing our concern with schistosomiasis and, at the same time, willfully ignoring the appeals of the TV messages.

Only about half of our informants could name the two forms of schistosomiasis, *ma'wiya* (*S. mansoni*) and *bawliya* (*S. haematobium*). Few of them were aware that hematuria (blood in the urine) was a symptom of only one form of the disease (*S. haematobium*), or that the two types required different diagnostic tests.

Forty percent of the residents of al-Garda and al-Salamuniya whom we interviewed (119 of 296) said that they had once been, or were currently infected. However, 58% of them did not know with which type they had been infected. Given that both types are treated with praziquantel, it is possible that staff at the rural health unit had never told them. Only 10 of the 118 (0.8%) who responded to this question claimed that they had had *ma'wiya (S. mansoni)*, compared to the 59 who said they had *bawliya* (*S. haematobium)*; three said they had had both. As we shall see in the next chapter, the last two responses reflect the fact that staff at RHUs mostly tested urine specimens to identify *S. haematobium*, even though, by the early '90s, *S. mansoni* had replaced it as the dominant form of schistosomiasis in the Nile delta.

Given this bias toward *S. haematobium* at the health facility, it was not surprising that most of what people in al-Garda and al-Salamuniya knew about schistosomiasis actually referred to *S. haematobium*. Seventy percent of the local adults interviewed said that blood in the urine was a common symptom of schistosomiasis. In addition, many of them mentioned symptoms such as "weakness," "paleness," "yellowness," and stomach cramps.

In the case of *S. mansoni*, its symptoms, which include blood in the stools, bloody diarrhea, and stomachache, may also be signs of other intestinal problems. Thus, it is more difficult for lay people to recognize *S. mansoni* than *S. haematobium*.

In the study communities, almost everyone knew that schistosomiasis could be diagnosed and treated at the local health facility. Ninety-four percent of the people we interviewed (278 out of 296) also knew that the infection could be treated by taking pills. The other 6% stated that pills and/or injections were required. This statement suggests that either they (or their parents) had known about, or been treated with, tartar emetic injections which had been standard treatment between 1920 and the early 1960s.

When we compared our findings in Munufiya briefly with those of researchers studying newly settled areas in Ismailia governorate, on either side of the Suez Canal, we found that Ismailia villagers identified symptoms of *bilharzia* or *harzi*, such as hematuria, nausea, and burning urination. Most of these symptoms were more closely related to *S. haematobium* than to *S. mansoni*. Ismailia villagers also identified "saltiness"—a collection of complaints including "burning urination," "kidney" and "tiredness"—as symptoms of schistosomiasis. What linked "saltiness" in the respondents' minds was a single feature, the saltiness of the surface water used both for irrigating crops and for daily domestic needs (Mehanna et al. 1997). Here (in contrast to the situation in our two Munufiya study communities), a single term was used to identify a condition that had no biomedical equivalent.

Schistosomiasis as a "dangerous" illness

Given that most people in al-Garda and al-Salamuniya knew that they could be diagnosed and treated for schistosomiasis, what did the 80% of survey respondents mean when they said it was a "dangerous" illness, *marad khatir?* Some suggested that it was "dangerous" because it had actually touched them, or family members; it was not something that only happened to other people. Others assumed it must be "dangerous" because so much attention was given to it on TV. We were also aware that some people told us it was "dangerous" because they knew that we had come to their community to study schistosomiasis, and so we, too, must consider it "dangerous" and that it was polite for them to consider it so as well.

One aspect of the "dangerous" nature of the illness was missing from villagers' discussions. When explaining to us why they should seek treatment for schistosomiasis, they mentioned that it could cause kidney failure, cancer of the bladder, and barrenness. But they did not think that they would be affected by these conditions if they took the pills to cure schistosomiasis. Perhaps they thought that all biomedical treatments necessarily worked; they were not aware that the drug did not always produce a cure (treatment failures for *S. mansoni* have been estimated at 20 to 40%, depending on local conditions). Villagers said that if they went back to the canal, they could simply go to the health center for another dose of pills. Here, then, the local concept of schistosomiasis as a "dangerous" disease was rather different from that of biomedicine. From their point of view, local residents were not being inconsistent when they told interviewers that schistosomiasis was a "dangerous" disease, while at the same time demonstrating an apparent lack of concern about actually contracting it.

Sources of information about schistosomiasis

When praziquantel was first made available in 1988, the Ministry of Health used TV messages to increase people's awareness of the causes of schistosomiasis, and its new treatment. Four years later, when we conducted our surveys and held formal and informal discussions with people in al-Garda and al-Salamuniya, most men and women claimed that their knowledge of the illness came from TV. As we have seen, most households in al-Garda and al-Salamuniya owned a TV. People who didn't could watch it at a neighbor's house, and men watched it at the coffee shop.

The situation of women was of particular interest, given that in most households women are the first providers of care for family members. In a survey about knowledge of schistosomiasis, 87% of 77 senior women interviewed told us that they received information about the disease from TV. As shown in table 10.2, just over a third said that TV was their only source of information, and a slightly smaller proportion said that they supplemented information from TV with information from relatives and neighbors—emphasizing the role of women's informal networking about illness in their families. Very few women said that their information had come from the local health unit or from talking to the doctor at a public or private health clinic.

Table 10.2: Source of women's knowledge of schistosomiasis

TV only	27	35%
TV, relatives and neighbors	25	32%
TV, relatives, had schistosomiasis	7	
TV and doctor	7	
TV and school	1	
Total mentioning TV	67	87%
From the doctor	7	
From school	2	
From the health unit	1	

N = 77 of 80 senior women

Source: household survey, 1993

People in both al-Garda and al-Salamuniya said that TV messages had warned them that canals were a source of infection. It also told them that they should: "Put your back to the canal." But as we have seen, this injunction had no connection with the real world of the everyday as understood by local people, so they simply ignored it. However, they were not unwilling to accept the TV instruction to go to a health unit for diagnosis and treatment if they thought they had picked up the disease. In a later chapter, we will explore in more detail the role played by these TV messages in health education.

Deciding where to go for treatment

Given that almost everyone in our two study communities knew that free diagnosis and treatment for schistosomiasis was available at their local health center, it is of some interest to explore what actual steps people took once they suspected they were infected or at least "at risk." By and large, adults made the decision about whether or not to seek testing independently of other family members. In doing so, however, women based their decisions on assessments of their situation that were rather different from those of men.

Our discussions with mothers showed that they relied on the school screening program to identify any of their school-age children who might be infected. As they trusted the efficiency of the authorities to conduct these screenings regularly, most mothers felt that they did not have to take any other action.

The situation was more complex for residents who were past school age. In a survey of 226 adults in 80 households in both villages, over 50% of adults claimed that at some time or another they had been tested and found positive for schistosomiasis (119 of 226 adults, 53%). We asked them where they had been tested. Many al-Salamuniya residents said they had been tested at the local rural health unit.

Table 10.3: Sources of diagnosis and treatment for schistosomiasis, 1993

RHU	20
Public hospital in Shebbin al-Kom	26
School testing	24
Ancylostomiasis unit	8
Private doctor	22
Other public hospital	4
Army hospital	3
Abroad	3
Tested by the project	8
No response	1

N = 119 (those tested positive and treated)

Source: household survey, 1993

A few people from al-Garda, where there was no health unit in the village, went to the nearest RHU. Other adults in al-Garda told us they had been to the public hospital in Shebbin al-Kom or to the ancylostomiasis (hookworm) unit (now defunct), as shown in table 10.3. They could easily reach these facilities by the minibuses that regularly traveled between al-Garda and Shebbin al-Kom. This journey was actually more convenient than going the slightly shorter distance to the RHU on the far side of the mother village. Of the varied sources used by these respondents in al-Garda and al-Salamuniya, and excluding those tested by our project and by health authorities overseas, we find that 82 of 107 respondents (77%) attended free public health services, indicating the predominant role of these services in the treatment of schistosomiasis.

However, our team soon discovered that gender made a real difference in searching for health care. The free public diagnostic and treatment facilities appeared to be easily accessible to all villagers—within walking distance of most houses in al-Salamuniya, and a short minibus ride for villagers in al-Garda. However, when we were planning the epidemiological survey for al-Garda (as there was no RHU in the village), women were very reluctant to leave the village for testing. They said that their many responsibilities—domestic and economic tasks, and child care—made it difficult for them to find the time needed for the journey, even if the research team provided transport. Thus, we decided to rent an apartment in al-Garda as a base for research activities, including testing and treatment.

In general, we found that women in al-Garda and al-Salamuniya (regardless of their domestic and other responsibilities) were accustomed to ignoring their own health needs and carrying on their daily life even if they felt ill. When they did eventually go to the health center, it was rarely because they feared that they had schistosomiasis. Fatma, an uneducated mother of young children said: "We do not know if we have schistosomiasis. We only go for analysis if we complain of another illness."

In al-Salamuniya, the Community Development Association clinic, part government-supported and part NGO, was able to provide diagnosis for schistosomiasis, which was otherwise usually only available in rural areas at MOH centers. Though testing and treatment for schistosomiasis at this clinic cost about 15 pounds Egyptian at the time of our study, people in full-time employment found its evening opening hours fit in with their busy schedules, and accordingly used its services.

Although private doctors in al-Garda and al-Salamuniya did not consider it cost effective to equip their offices with the laboratory equipment and specialized staff needed to carry out the parasitological diagnosis of schistosomiasis, they could refer patients they suspected of having schistosomiasis to the RHU. However, they were more likely to suggest that, as praziquantel was available at the local pharmacy and had no side effects, the patient should simply purchase a supply and treat themselves. Some local farmers, who knew they were at risk because they visited the fields regularly, told us that they purchased praziquantel from a pharmacy every six months or so, without visiting the RHU or a private doctor.

In both al-Garda and al-Salamuniya there were many "informal" health providers who provided simple wound treatment and injections, and recommended medications that could be purchased at a stall in the market or in the pharmacy. Some of these providers had formal biomedical training and worked, or had once worked, at local health facilities, while others had no formal biomedical training and could be said to be part of a "folk" tradition. What was distinctive about schistosomiasis was that very few villagers stated that they used any local medicine once they suspected that they had the disease. For schistosomiasis, villagers clearly recognized that they needed to look for diagnosis and treatment within the biomedical sphere.

The "medicalization" of schistosomiasis

In al-Garda and al-Salamuniya the dialogue about the disease entity known locally as *bilharzia*, and how it should be treated, appeared to incorporate many biomedical concepts and assumptions affecting how people recognize that they have a disease, and what they decide to do about it. Before we discuss this in more detail, it is useful to look, by way of contrast, at the minimal (or even non-existent) local dialogue about reproductive health problems.

In some rural areas of Egypt in the late 1980s and 1990s, women viewed reproductive tract conditions such as vaginal discharge and even uterine prolapse as "normal" occurrences that did not require medical attention. They were also deterred from seeking treatment because of the stigma attached to any reproductive health problem. Moreover, there was little discussion among husbands and wives, and among women, about such issues. Even if there had been a village dialogue, there was no way this could connect to a medical dialogue at the local level. The few public or private reproductive health services that were available locally provided treatment only for infertility and for the delivery of babies. There was no health education to draw women's attention to the importance of reproductive health issues, and to encourage them to seek treatment (Khattab et al. 1999; Talaat 2001). In other words, women's reproductive health (except for issues relating to child birth and infertility) has yet to become medicalized.

In considering the extent to which schistosomiasis in our study communities has become medicalized, we should look at studies in other areas of the country, where, for one reason or another, local knowledge was rather different and the disease was less medicalized. For example, in the early 1990s in rural Ismailia, close to the Suez Canal, some villagers mentioned home remedies for schistosomiasis. One was made up of the dried skin of pomegranate, boiled with cumin and an herb known as *shiikh*. Ismailia people knew, from TV and other sources, that testing and treatment was available at primary health care facilities, but these were often distant from the small, scattered settlements in which most people lived (Mehanna et al. 1997). In this respect, Ismailia (and other newly settled areas on the fringes of the Delta) differed from the densely settled irrigated lands of the central Delta (including our study villages) that had a more homogeneous population and more health facilities.

Fifteen years earlier, a study of *S. haematobium* in Upper Egypt found that people used both biomedical and "traditional" treatment. Of 162 males ten years old and above who said that they had been treated for schistosomiasis, 52 (46%) used only oral drugs and/or injections, 43 (38%) used drugs and herbs, and 17 (15%) used only herbs. The herbs mentioned included an infusion of *damsissa*, *Ambrosia maritima*, a plant that grows on the canal banks and in the fields, and is also used to kill snails. However, informants in that study, which predated the introduction of praziquantel in the national control program, mentioned that herbal treatment was becoming less common as oral drugs were becoming more available (Kloos et al. 1982)

In most parts of rural Egypt, this process of "medicalization" (working toward the acceptance of western medicine as the only appropriate treatment of schistosomiasis) began during the early 1920s, when the treatment with tartar emetic rapidly became common in the densely settled rural areas of the Nile valley and Delta. In this government sponsored program, special treatment camps, primary health care facilities or mobile clinics provided people who were found positive with a series of injections over a four week period. Indeed, schistosomiasis control was the on-going biomedical intervention that was most familiar to all rural Egyptians. Most programs, such as those for the control of cholera and malaria, were only activated during an epidemic (Gallagher 1993: chs. 2 and 7).

Although private and charitable health provisions were available in some rural areas, free and accessible public health services did not become widely available until after the Free Officers Revolution of 1952. These facilities continued the earlier established public schistosomiasis program. Private medical services in rural areas, expanding in the 1970s, could provide a clinical diagnosis of schistosomiasis, but rarely had facilities for the parasitological tests needed for a definitive diagnosis of the infection. Thus private doctors and clinics never played more than a minor role in the diagnosis and treatment of the disease.

The introduction of praziquantel into the government schistosomiasis control program in 1988 and the extension of TV messages about the disease gave considerable impetus to the process of "medicalization" of the disease. In the two study communities, al-Garda and al-Salamuniya, residents knew that they could go to health units for diagnosis and treatment, and that their children would be reached through the school based testing program, and they acted accordingly. This "medicalization" of schistosomiasis suggests, at a minimum, that both health providers and patients shared an understanding of the purpose of the patients' visit, what constituted a diagnosis (a urine and/or stool examination), and what constituted treatment (swallowing tablets).

One of our team member's earlier research projects on informal and formal health services in other communities in Munufiya governorate suggested that, in general, people had an eclectic approach to health care and treatment. They had access to both formal and informal care, and their choice depended on the way they defined the illness and what they considered was the most appropriate source of treatment (Asaad and El Katsha 1980). If one accepts that this view operated in our study communities in the early 1990s (and there is no reason to believe that it did not), it is clear that local people placed schistosomiasis firmly in the category of illnesses that only responded to modern western drug treatment, specifically praziquantel.

As we have seen, these shared perceptions might not have extended to a shared assessment of the "danger" posed by disease. The medical doctor, thanks to his or her professional training, knew that a particular individual might not be cured by praziquantel. On the other hand, individuals from al-Garda and al-Salamuniya took it for granted that praziquantel would cure their own infection. Some purchased the drug on the market and treated themselves. The existence of some kind of a dialogue about schistosomiasis at the community level simplifies our study of what actually happens when a villager goes to a health facility, for it assumes some common understanding of the interactions involved. We will explore this process in the next chapter.

11

Screening and Treatment Programs in Practice

Testing and treatment at the village level—the setting

In our study villages, residents viewed schistosomiasis as a disease best diagnosed and treated through the biomedical system, and largely through public facilities under the aegis of the MOH Schistosomiasis Control Program. In this chapter, we will focus on service delivery at the local level through this Program. Then, having observed the system in practice, we will turn our attention to strategies to make it even more effective. As all Egyptian MOH health units were administered in the same way, these strategies could be replicated at other health centers in the country. Our research also suggests that the problems we noted in Munufiya governorate are likely to be similar to those experienced elsewhere. The *Egypt Human Development Report, 1997/98,* issued by the Institute of National Planning in Cairo, pinpointed some of these organizational problems, such as a lack of coordination between different parts of the health system and a failure to make optimum use of available resources (EHDR 1998: 61–66).

Many doctors and nurses who practice at MOH facilities in our study villages (as elsewhere in Egypt) provide private services to local residents later in the day and in the evening. Thus, any activities that lead to improvements in staff knowledge about schistosomiasis and in the quality of services will directly benefit practitioners in both the public and private sectors. Overall, however, when it came to seeking treatment for the disease, more people attended public health facilities than private services for schistosomiasis treatment.

During 1992 and 1993 we worked with staff at the al-Salamuniya Rural Health Unit to identify screening and treatment practices, and later designed and tested strategies to upgrade these procedures. We also worked closely with staff from the RHU in al-Garda's mother village, who assisted us in the base we had set up in al-Garda itself. In 1994, after we had been requested to monitor the establishment of a public laundry in the village of Baguriya, we collaborated with local staff to assess schistosomiasis treatment practices at the RHU in that community. Throughout this time, during both formative research and "action," we also worked in collaboration with MOH staff at the district level.

When we first came to the al-Salamuniya RHU in 1991, we found that the most frequently used services were the diagnosis and treatment of schistosomiasis, services for mothers and children, and an outpatient clinic dealing with minor illnesses. General services for mothers and children were provided every morning during the working week and included the provision of Oral Rehydration Salts (ORS) for infants who had become dehydrated because of diarrhea. Immunizations for childhood diseases and tetanus immunizations for women were usually available two days a week. The outpatient clinic was open every day; the charge book indicated that the average daily attendance was between 15 and 30 patients. The health unit was open from 8 a.m. to 2 p.m. every day except Friday and public holidays; most patients attended between 9 and midday.

Services at the rural health units were available free to children referred to the RHU by the school nurse and for the school-based schistosomiasis program. Mothers coming for immunizations and for maternal care for their infants were also treated free. All other patients attended the outpatients clinic, for which they were charged a flat rate fee of 25 piasters for a consultation (about 8 cents US). This fee was increased to LE 1.10 in 1997. All drugs, such as praziquantel, were provided free. The general pattern of activities at al-Salamuniya RHU was fairly typical of rural health units in Munufiya governorate.

Health unit staff and responsibilities

The health facility at al-Salamuniya, in common with others in Munufiya governorate, did not suffer from a shortage of staff; there were many qualified people who wanted jobs in their home area. In fact there may have been more secretaries in al-Salamuniya RHU than were strictly necessary, as shown in table 11.1. These figures also show that, except for doctors and custodians, jobs were gender segregated.

Although the stated policy of the Ministry of Health was to transfer physicians frequently, this policy was not strictly followed. In al-Salamuniya, the man who was the supervising physician and the second physician, who was his wife, held these positions throughout the whole of our five year research period. When we began working in Baguriya, the woman superintending physician had been in place for over 15 years.

Table 11.1: Staff at al-Salamuniya and Baguriya Rural Health Units

	al-Salamuniya 1992			Baguriya 1994		
	Male	Female	Total	Male	Female	Total
Physician	1	1	2	—	1	1
Nurses	—	10	10	—	9	9
Dentist	—	—	—	—	1	1
Technicians	1	—	1	1	—	1
Secretaries	—	15	15	—	5	5
Custodians	2	2	4	2	2	4
Health Inspector	1	—	1	1	—	1
Asst. H. I.	—	—	—	1	—	1
Total			33			23

The same continuity of personnel was in evidence among the nurses (all women) and support staff. Many of them had lived for a long time in the community they served and provided a valuable link between the RHU and local people.

As part of his or her assigned role, the physician in charge of a RHU was responsible for all schistosomiasis control activities, including maintaining the lab and supervising the lab technicians who performed the parasitological tests. The physician was also responsible for administering praziquantel to patients and school children who had tested positive, and for referring patients with serious conditions to a government hospital outside the village.

In addition to schistosomiasis-related tasks, the physician in charge was responsible for all the activities of health unit staff. He or she was expected to maintain the unit's financial and attendance records, and ensure that the information required by the district health office was sent in on time. He/she was also responsible for ordering supplies for the daily operation of the health facility and drugs for the pharmacy. Burdened with all these responsibilities, it is not surprising that the physicians in the health units with which we were associated told us that they did not have time to do more than glance at patients and their health results; they certainly could not take on more responsibilities.

Despite the importance of schistosomiasis as a threat to rural health, no specific nurse or nurses at al-Salamuniya or Baguriya RHU had overall responsibility for schistosomiasis control activities. When, for example, a flood of children descended on the health unit for screening, the physician in charge might ask any of the nurses to assist. Thus, many of the nurses at al-Salamuniya had some experience with schistosomiasis, but none were specialists. In contrast, certain nurses were regularly assigned specific responsibilities in the area of immunizations or maternal care.

During the time of our research project, the maternity nurses visiting mothers in the village who had recently given birth were the only nurses who regularly left the clinic to make home visits. No nurses were assigned to make home visits to ensure that people who had been given praziquantel would return to the health unit three months after treatment for retesting. This waiting period is necessary as praziquantel does not kill immature worms, and eggs do not appear in the feces or urine until 30 to 40 days after infection. Because of this lack of follow-up, many of the 20 to 40% of *S. mansoni* patients for whom praziquantel had not been effective slipped through the net, as well as those who became reinfected during this time.

During the early and mid-1990s at al-Salamuniya and Baguriya, no health unit staff had a specific obligation to carry out health education. Instead, responsibility for this important aspect of preventive public health lay with the Health Education Unit at the district Department of Health. Periodically, staff of this unit conducted campaigns at the local level, related to specific programs, such as immunization, oral rehydration, family planning, and schistosomiasis. This assignment of responsibility for health education to the district rather than the local level (where infection takes place) in part explains why staff at rural health units did not find it in their mandate to provide clients and patients with information about how to avoid becoming infected with schistosomiasis.

School nurses (in the RHUs where we worked they were all women) were posted to the school rather than the health unit, although they were MOH employees. They were required to liaise with the physician in charge of the health unit to arrange for the periodic schistosomiasis screening of all children in their school. However, they were not expected to tell children about schistosomiasis, nor did they have the necessary knowledge to do so.

The position of the snail inspector in the administrative structure was also less than ideal. Based at the district health office, the snail inspector was supposed to tell staff at the health unit if there were any infected snails in village canals, and when these sites would be treated. However, we noted that the snail inspectors did not report their findings to the health units, nor did they realize that their findings would be of real importance in preventing further infection.

In the health units at al-Salamuniya and Baguriya, the sole technician (male) was responsible for preparing and examining patients' urine and stool specimens, and for supervising the secretary (female) who recorded the results of his tests. This secretary sent the records to the physician in charge, who (as required) sent them on to the Endemic Diseases Unit in the governorate, with a copy to the *markaz* (district) Department of Health. The technicians were also supposed to take a 10% sample of all village households for screening, and record the results. We found that they did this mainly during MOH campaigns organized from the district headquarters.

The custodians in each of the two health units kept the place clean and tidy and fetched and carried for the doctors and nurses. However, over and beyond this, they frequently served as the first point of contact between an incoming patient and the health unit's professional staff. These newly arrived patients and other people who had been sitting around waiting for attention often asked a custodian's advice about schistosomiasis and other health problems. When physicians or nurses saw patients they were usually too rushed to answer patients' questions or provide guidance on prevention.

Our overall impression of the organization of the tasks related to schistosomiasis at the two health units was of a lack of coordination. This problem was in part due to the absence of official written job descriptions. The highly centralized system, which also depended on horizontal coordination at certain levels, left staff uncertain of their position and responsibilities. As a result, staff, when they described to us what they were expected to do, saw their tasks as a series of disconnected activities rather than part of a coherent process.

With regard to schistosomiasis, no single person in the health unit, not even the physician in charge, had an overall view of the tasks involved. The doctor gave patients pills but held no one responsible for finding out how these people had become infected, so that preventive measures could be undertaken. No staff member was able to follow through three months after treatment to make sure that it had been effective. A similar lack of follow-through attended the work of the snail specialists; they were not required to report their finding to the RHU staff so that the latter could warn villagers about sites where there were infected snails, or where chemicals had been put in the canal.

On the basis of these findings, and with the support of the Ministry of Health, the research team decided that our first task should be to hold training workshops on schistosomiasis for all health unit staff. Our intent was to instill in them the understanding that schistosomiasis control and prevention must be a unified and interrelated process, if it was to succeed.

Staff knowledge and perceptions of schistosomiasis

Working in cooperation with the physician in charge of each rural health unit, we held workshops for nurses, technicians, and other staff who directly interacted with patients visiting the unit. In order to plan the workshops, we first held some focus group discussions, interviews, and informal discussions with health unit staff, and found that their knowledge of schistosomiasis was somewhat incomplete. Although most of them knew that schistosomiasis was associated in some way with exposure to canal water, their knowledge of the transmission cycle was limited. They seemed not to be clearly aware that two distinct kinds of schistosomiasis were found in their area. Indeed, most of them equated schistosomiasis with *S. haematobium*. They said that the main symptom of schistosomiasis was hematuria, blood in the urine, but they did not identify this symptom with a particular form of schistosomiasis. And although the staff all agreed that it was a "dangerous" disease, they were unable to say clearly why this was so. They were aware of its possible fatal consequences but, like lay people in the community (who were unaware that praziquantel did not always cure them), they thought these could be avoided if treatment was provided. However, the main reason why RHU staff considered it "dangerous" was because it was regarded as important by the MOH and was a major aspect of their day to day work.

None of the nurses felt that, under the current conditions, they had a specific role in the schistosomiasis program. As they explained it, only the technicians and the school health nurses had such a role. As for health education about schistosomiasis, they agreed that this was the duty of the doctor (who administered the praziquantel) and of the technicians. The majority of the nurses did agree, however, that perspectives should be broadened and that information about the disease should be incorporated into various RHU programs, such as the Mother and Child Program and Immunizations.

Only two of the 12 staff interviewed told us that they had received specific information about the schistosomiasis life cycle during their training. Staff recognized that their information about schistosomiasis was sketchy and they expressed an interest in learning more about the disease and expanding their roles and expertise in this area.

Parasitological testing in practice

When we arrived on the scene, staff at the health facilities serving al-Salamuniya and al-Garda knew that they were expected to follow the standard MOH protocols for diagnostic testing. This required that all patients should provide urine specimens to identify the ova of *S. haematobium* and stool specimens to identify the ova of *S. mansoni*. Recent research had shown that, over the past three decades, *S. mansoni* had largely replaced *S. haematobium* in the Delta (Cline et al. 1989). However, our discussions with health staff, and observations of practices, soon showed that staff knowledge and practices had yet to catch up with this reality.

This time lag was first brought to our attention when we observed that, in their day to day work, technicians at the al-Salamuniya RHU, and those from the nearby RHU who worked for the research project in al-Garda, focused on the collection, preparation, and examination of urine samples, rather than on both urine and stool samples. They knew how to perform the sedimentation test on stools, as required by the MOH, and had the necessary equipment. However, when questioned about why they did not carry out stool examinations on a regular basis, they stated that they preferred to test urine specimens, as testing stool specimens was a messy and unpleasant procedure. They also stated that patients were sometimes reticent to furnish samples on demand, and that neither the doctor in charge, nor the nurses, had warned patients that they would be expected to provide stool specimens. Overall, we came away with the impression that RHU staff were not aware of the importance of *S. mansoni* and the need to collect stool specimens (El Katsha and Watts 1995 A; 1995 B).

To assess the extent of the neglect of *S. mansoni* in al-Salamuniya, we compared the results of our epidemiological surveys with the health unit records. In our pilot survey carried out early in 1992, less than 1% of those tested were positive for *S. haematobium* (thereafter, we conducted no further tests for *S. haematobium*). Our main survey, conducted later that year, found that 25% of those tested were positive for *S. mansoni*. In contrast, the record book for the al-Salamuniya health unit for June–July 1991 indicated that only 5% of those tested were found positive for *S. mansoni* and 9% for *S. haematobium*; this strongly suggested that in most cases, only urine specimens were examined.

In June and July of the following year (1992), the RHU records identified that 40% of patients had *S. mansoni* and less than 1% had *S. haematobium*. This dramatic increase in the detection of *S. mansoni* had occurred after the technician had attended our local workshop on schistosomiasis and a training course on the analysis of stool specimens. The discrepancy between our findings (25% prevalence) and the RHU records (40% of those tested) had probably come about because the al-Salamuniya doctor (by his own admission) had only authorized testing for people who seemed to show some signs or symptoms of schistosomiasis. He had not referred all patients for testing; in short, this sample was skewed.

Looking at the broader picture in the district in which al-Salamuniya is found, official returns for the years 1985 to 1991 suggested that testing for *S. mansoni* only increased gradually. *S. mansoni* comprised 16–17% of all schistosomiasis cases in 1985, 1986, and 1987, increasing to around half of all cases in 1990 and 1991; denominator figures, indicating the actual number tested, were not available in these district level records. Yet, in the Munufiya governorate the EPI 1, 2, 3 survey in 1992 found that *S. haematobium* was being transmitted in only *one* of the 27 communities studied (Abdel-Wahab et al. 2000 B). Thus, it appears that the pattern of testing in al-Salamuniya was similar to that of other health units in the district and most likely also in the governorate as a whole. This resulted in the serious underreporting of *S. mansoni* in official MOH records.

The need to test stool samples for *S. mansoni* is not limited to the Delta. As pointed out in chapter 5, recent studies have shown that this form of schistosomiasis is now being locally transmitted in the Nile valley south of Cairo (Abdel-Wahab et al. 1993). In the late 1990s, in a village in Giza governorate, just south of Cairo, the prevalence rate for *S. mansoni* was approximately four times that for *S. haematobium*. Alarmed by these findings, the Giza researchers and the health staff who worked with them also appealed for training and education programs to be directed toward testing for *S. mansoni* (Talaat et al. 1999 B).

Upgrading services at the Rural Health Unit: Training

Health staff's limited knowledge of schistosomiasis, and the neglect of stool testing at the RHUs alerted us to the need for staff training. The MOH had provided technicians with some training on parasitological testing techniques. However, they failed to provide technicians or any other health unit staff with information about schistosomiasis. The situation was summed up by a nurse who assured us that: "Training for schistosomiasis is not a regular concern. Topics for training are provided according to the grants available to the Ministry."

Accordingly, after we became aware of this situation, our first activity in al-Garda and al-Salamuniya (and later in Baguriya) was to hold an interactive training workshop for all health staff who had an actual or potential role in the control program. To overcome the fragmented approach which was so much in evidence, we began the workshop by emphasizing the need to work as a team, and to adopt an overall view of the problem and of the ways it might be solved. In the course of this training workshop our team members built up good rapport with all the health unit staff, from the physician in charge to the custodians. The dialogue established there continued throughout the research period.

The training workshop was both participatory and interactive. The format we adopted encouraged open, informal interaction between trainers and RHU staff, with discussions and questions in small groups. This represented a move away from the didactic, lecture format and its emphasis on rote learning (familiar to participants since their days in primary school), toward a participatory exploration of issues identified by participants as well as by trainers.

The researchers also arranged for the technicians to receive training at the MOH Center for Field and Applied Research (CFAR) in Qalyub. There they were introduced to the Kato-Katz technique that was to be used in our epidemiological surveys, following the protocol established for studies supported by the Schistosomiasis Research Project. Members of our research team and, later, MOH district level staff monitored the technicians' performance and provided them with in-service training.

The record-keeping system

Because schistosomiasis is a disease that is related to human behavior and to environmental conditions in a given locality, people who are directly involved in controlling it at the local level must have information about the situation they are addressing. However, as of the early and mid-1990s, information of this sort was collected solely for the purpose of sending it up to the Ministry of Health in Cairo, via the district and governorate MOH departments. The MOH in Cairo uses this information for planning its own programs and might also send it to organizations such as the Information and Decision Support Center (IDSC) in Cairo which compiles data for national policy and planning.

In this centralized system, the movement of information is *from the periphery* to the center, and thus there is little, if any, feedback to the local level. During our period of research, as far as we could tell, no provision was made for using information collected at the RHUs in al-Salamuniya or in Baguriya for local level decision-making. This meant that the local people who generated the statistics could not see the benefits of what they were doing. Accordingly, their morale was low, and they were not motivated to ensure that the figures they collected were accurate.

When we began looking at the recording system at al-Salamuniya Health Unit in mid-1993, we found two sets of relevant medical records. The first was a general register containing the names of all the patients who had come into the health unit each day for one or another of the endemic diseases, including schistosomiasis. The second was a form listing information taken from the daily register and then sent on to the district Department of Health every month. Clinic staff stressed that for them, it was important to make sure that the reports to the district were correct and met the MOH requirements. Therefore they had to keep clear records, indicating, in the case of schistosomiasis, how many patients they tested, the results of the tests, and the number of praziquantel tablets given out.

Looking at the situation further up the hierarchy, at the district level, as of the early 1990s we found aggregated records which listed the number and sex of those who tested positive for *S. mansoni* and *S. haematobium* for all the RHUs in the district. However, to be useful for planning purposes the information needed to include a denominator—the total number of individuals tested—as well as an indication of the sex of those tested and those found positive. Data on gender is important, as a gender imbalance or a change in the figures for males and females could indicate a failure to reach a certain group, such as school age girls who were not attending school, or young women. As of the early 1990s, the statistical information found at the district level (collated from information sent in from the local health units) was almost impossible to use in any meaningful way in planning at that level. Further aggregation of this data at the national level meant that there was no information on gender in national statistics.

Another key failure of the record keeping system at the level of the RHU was the absence of any way to follow up treated patients to remind them to return three months after treatment in order to check whether or not they were clear of the disease. Patients not cleared of infection might infect others if their urine or feces reached a canal.

Improving the local records

After we had held extensive discussions with local and district level MOH staff, we revised the daily register form and tested it in the al-Salamuniya RHU between February and June 1994. The main objective of the new form was to ensure that treated patients were followed up three months later. A column was added for this purpose, on which the results of the follow-up visit were to be recorded. At the bottom of the form was a reminder to the technician (responsible for supervising the secretary who entered the information on the form) to test each patient for *both S. mansoni* and *S. haematobium* and to record the sex of the patient. The technician was also reminded to insist that the patient come back in three months for re-testing.

We also thought it worthwhile to give each departing patient a card telling them to come back in three months. The card also contained messages reminding them that even if they were cured by praziquantel they were at risk of becoming infected again if they continued to use the canals. The message also warned them that there were two types of schistosomiasis.

The card system did not appear to work out well in practice. Between February and June 1994, only two people returned with their cards and at the right time. However, the cards did create a new awareness of disease danger. Many people came back well before the end of the three month period, usually without their cards. They were worried that they had been infected again, and insisted that they should be given another dose of praziquantel.

The individual cards and the new recording forms should be seen in the context of the whole recording system at the RHU. Because staff still needed to maintain the existing record book for endemic diseases, they found that the new form, introduced for a short time on an experimental basis, was simply an additional burden. Also, it took them extra time to look back in the register to find the patient's name and to record that he or she had come in again. They found the existing system to be more straightforward, as it simply required a separate entry each time a patient came in, with no attempt to correlate this entry with earlier ones.

From our limited intervention, it was clear that for any new recording system to be effective, the users, both health staff and villagers, should be properly informed about its purpose. It is not enough simply to introduce a new recording system (or a new diagnostic procedure or treatment) into the health care setting. Health providers need training and patients need clear instructions if such an intervention is to be effective. Records need to have a clear purpose, in this case the recording of test results by sex and age, treatment, and follow-up. To make full use of any upgraded records, local health unit staff and MOH staff at the district and governorate level should review the information and use it as a guide for action.

The school-based screening program

When praziquantel was introduced into the schistosomiasis control program in 1988, the Ministry of Health recognized that the promise of a high cure rate made it essential to continue, and improve, the existing provisions for school-based screening and treatment. Epidemiological studies in Egypt (and elsewhere) had shown that, in general, children under the age of 15 have higher intensity and higher reinfection rates than the rest of the population. Intensity and infection rates were usually lower for older people who had more limited exposure, and who had acquired some immunity with age and with exposure (see chapter 3).

Our own epidemiological studies in al-Garda and al-Salamuniya gave a mixed picture of schistosomiasis infection among school-age children. Our 1992 survey showed that only 5% of al-Garda children aged 5 to 14, and 15% in al-Salamuniya, were infected; in both communities these proportions were smaller than for the total population. Our second survey (in 1993), showed that just under one third of all new infections in the two communities (22 of 61 cases) were among school-age children. But given that the 5–14 cohort constituted nearly a third of the total population tested, this proved very little except that school children in both villages were continuing to be infected.

Our main concern, however, was to look at the school-based program's testing practices for *S. mansoni,* with a view to upgrading them. The health unit records showed that, in the first six months of 1992, staff had detected a *S. mansoni* rate among school children of 1%, compared to 5% among adult patients. This suggested that staff had more problems with school testing procedures for *S. mansoni* than with those used for adults who came in individually. We recognized that a well-run school-based program would test most of the children in primary and preparatory schools (aged about 6 to 14), given that, in our study communities, 95% of all boys and 91% of all girls in this age group were actually attending school (El Katsha and Watts 1998).

As our study was beginning in 1991, direct responsibility for overlooking the health needs of the country's 16 million school children was in the process of being transferred— in Munufiya governorate and nation wide— from the Ministry of Health to a subsidiary organization, the Health Insurance Organization (HIO). This represented a considerable expansion of activities for the HIO (which had formerly mainly provided services for government and private sector employees) and entailed a major reorganization of school health programs (EHDR 1998: 52, 58). During the time of our research in Munufiya governorate, schistosomiasis screening had not yet been transferred to HIO centers. Indeed, at that time, it was not certain when, or even if, MOH health facilities would actually cease providing screening for school children. Accordingly, we pressed forward with the planning of a revised screening strategy.

Our first task, early in 1993, was to describe the existing practice of school-based screening. When we came to consider how best to revise this strategy, it was important to note that *all* children were expected to go to the health unit to provide the specimens needed for parasitological analysis. As of early 1993, the school nurse (who was part of the school's establishment rather than that of the RHU) collaborated with health unit staff to arrange the times for testing. She told the children ahead of time that they were going to the RHU, but she did not give them any information about schistosomiasis or what to expect during the visit. Most important, the school nurse did not tell the children that they would be expected to produce both stool and urine specimens when they got to the health unit.

We saw for ourselves what happened when a class of fifty boys and girls arrived en masse at the al-Salamuniya Health Unit. Not having been forewarned to bring specimens, the children lined up to receive containers and were ordered to produce both a urine and a stool specimen. Since few of them were able to produce a stool specimen on demand, the technician had to make do with urine specimens. We could now clearly understand how it was that the al-Salamuniya RHU staff in the first half of 1992 had come up with a figure of 1% of school children infected with *S. mansoni*.

Our team also noted that the RHU staff were insensitive to both boys' and girls' need for privacy. Although girls and boys were assigned separate bathrooms, during the rest of the procedure they mixed indiscriminately. The names of the girls and boys who tested positive were read out in front of the all the assembled children, and they were then taken to the physician's office to be given praziquantel. Although the children regarded schistosomiasis as a common and inevitable infection and one not associated with stigma, they were upset by this public recognition of their own infection.

In view of these problems, it is not surprising that girls, who were more sensitive to violations of their privacy than were boys, often decided not to go to school on the day the screening was to take place; or, when the group left the school grounds, they surreptitiously returned home. The problems encountered by girls arose largely because the gender-segregated patterns of behavior accepted in the school were not carried over into the health center. As a matter of custom, social interaction in the playground was largely gender segregated, and in the classroom boys and girls sat on opposite sides of the room.

In addition to the less than complete coverage that resulted from these procedures, there were other difficulties. The sole technician at the al-Salamuniya RHU could not, on his own, prepare and examine 50 or more urine and stool specimens in one day: a well trained technician could process about thirty urine or stool samples in that time. Again, we understood why the health unit reported a *S. mansoni* rate for children of around 1% in early 1992. Also weakening the value of the school-based screening program that we found in operation upon our arrival was the absence of a three-month follow-up for children who had been treated with praziquantel.

Designing and testing a gender-sensitive screening strategy

In designing the new strategy which would be given a trial at the al-Salamuniya health unit, we used the core concept of action research: the active involvement of all participants, both at the RHU and at the primary school attended by the children who were to be tested. At the school, we met first with the school nurse and the teachers and told them about the threat posed by *S. mansoni* and its long-term impact on students' health. We pointed out that on most occasions the existing screening procedures did not detect this form of schistosomiasis and asked them to cooperate with us in devising a more effective procedure. At this stage, the school headmaster joined in and promised to do his utmost to help us; his contribution was invaluable as he was respected by teachers, parents, and children in the community. We also discussed the procedures for school children with the staff at the RHU, who made us fully aware of the problems they faced, given the limited daily capacity of the lab and the congestion and confusion caused when all the children came to the health unit. Then we held several meetings in which health unit staff, the school nurse, and teachers shared their ideas about how to improve the procedures.

In November and December 1993, we tested the revised strategy among 9 to 11 year old children enrolled in two year 4 and two year 5 classes in an al-Salamuniya primary school. On the day before the specimens were to be collected, the headmaster explained the new procedure to the children, giving basic information on schistosomiasis and stressing that each of them should provide his or her own stool specimen. Then we introduced our first major innovation, involving the distribution and collection of stool specimens. The school nurse and teachers provide the students with containers (clearly marked with the child's name) on the day before the testing was scheduled. Early in the following morning, the RHU custodians (two women and two men) went to students' houses to collect these containers and to bring them to the health unit for testing by the technician.

Our second innovation was to make sure that the technician had time to do his work properly. Rather than rushing through his work while the children crowded outside his door, he was given time to prepare and test specimens at his own pace. As soon as he had the results, they were sent to the school. The class teacher told each child privately if they were infected, or if they needed treatment for other parasitic infections, most commonly oxyuriasis (pinworm or thread worm) and ascariasis (an intestinal nematode). The school nurse then discreetly accompanied separate groups of boys and girls to the health unit for treatment. Using this revised strategy, it was found that the *S. mansoni* prevalence rate for the four classes of 9 to 11 year olds was 25%, compared to the 15% of 5 to 14 year olds who had been found infected in our 1992 epidemiological survey.

Several factors contributed to making the gender-sensitive strategy effective. Firstly, the prior distribution of containers for stool specimens meant that girls and boys could provide these in the privacy of their own homes, rather than being expected to provide them on demand, in the crowded setting of the RHU. Disruption at the health unit itself was avoided as only those children who required treatment needed to visit it.

This revised strategy required no extra staff, no extra equipment, and no extra expense (El Katsha and Watts 1998). Thus, it could easily be introduced on a permanent basis in all Rural Health Units. Improved staff training and cooperation between health unit staff, the school nurse, and teachers would be needed for the full benefits of this gender-sensitive strategy to be realized. Above all, it has to be approved by the MOH administration and incorporated into the system.

Reaching children who are not in school

In the Munufiya governorate study area, a very high proportion of school-age children attended school, and thus could be reached in a school-based screening program. However, it is well known that rates of school attendance are not uniform throughout Egypt. They are much lower in rural Upper Egypt, especially for girls. Though school enrollment is compulsory for all children between the ages of 6 and 14, formal *enrollment* is not the same as regular attendance in school. As of the mid-1990s, actual school attendance figures in Egypt's schools were not routinely available.

In 1996, Husein and colleagues published a study of 30,000 children in Upper Egypt between 6 and 15 years of age, using data collected by the EPI 1, 2, 3 survey of 1992. They reported that 81% of the girls found to have schistosomiasis were not attending school, compared to 41% of boys. This meant that more than eight out of ten infections among school-age girls, and four out of ten among boys, would have been missed in any testing program based on schools. Husein also found that on average, infected non-attending children had higher egg counts than did infected children attending school. This suggested that non-attending children actually experienced more morbidity, and had a greater potential for transmitting infection than did infected children who attended school (Husein et al. 1996).

These findings about the plight of non-attending children in Upper Egypt prompted that research group to explore ways of reaching such children in Fayoum governorate, 60 km. south of Cairo. This governorate had the lowest level of primary school enrollment for girls and for boys in Egypt, with only 41% of girls enrolled in school in 1986–87, compared to 79% of boys (Allen 1989: 53, 55). In a pilot project researchers tested almost 4,000 children in twenty Fayoum hamlets, *'izab*, none of them further than 3 km. from a school that provided treatment. Teachers volunteered to help contact families, and students were asked to recruit siblings and friends who did not attend school regularly.

In the twenty Fayoum settlements, the researchers found that levels of non-attendance ranged from 36% to 89.3% for girls, and between 17.8% and 61.9% for boys. And, in an in-depth study in two hamlets, many parents told researchers that girls did not need to be educated, but should stay at home and learn how to be good wives and mothers. Fathers, especially, expressed the belief that it was not good for girls and boys to mix, even in primary school. However, the girls interviewed had a very positive view of the value of school.

Working in these small Fayoum settlements, researchers were able to collect urine samples from 90.1% of the non-attending girls in the census list, and 85.6% of the boys. Overall, 64.8% of the girls, and 67.6% of the boys tested positive for *S. haematobium* (as *S. mansoni* rates were very low in the study settlements the screening only obtained urine specimens) (Talaat and Omar no date: Talaat et al. 1999 A).

The girls were especially appreciative of the efforts taken on their behalf by the researchers. They told the chief investigator that this was the first time any group had tried to help them in any way. This brings us back to our main point, that all of these young people in Fayoum who tested positive would have been missed by a school-based testing program, given that none of them were attending school.

Female genital schistosomiasis

During the 1990s, researchers in tropical Africa began to identify female genital schistosomiasis (FGS) as a health problem for women living in areas endemic for *S. haematobium*. They found that it was often overlooked because it is difficult to diagnose and physicians could confuse its symptoms with those of sexually transmitted diseases (Poggensee et al. 1999). In Egypt in 1997, in the first-ever study of FGS in a community, researchers found that half of women of reproductive age in a small hamlet in Fayoum governorate were affected (Talaat 2001). FGS due to *S. haematobium* occurs when eggs leave the urinary tract and move to the genital organs. The eggs, with their characteristic terminal spine, give rise to chronic inflammation in the surrounding genital tissue, and cause women considerable pain and discomfort. Although most cases of FGS have been found to be due to *S. haematobium* (as in Fayoum), hospital examinations elsewhere in Egypt have identified cases due to *S. mansoni*.

As women with FGS may not be excreting eggs, a routine examination of a urine or stool specimen would not necessarily indicate that they were currently infected with schistosomiasis. Thus, women suffering from FGS would not necessarily receive praziquantel, which can cure the early stage FGS lesions and get rid of schistosome eggs in the genital tract (Richter et al. 1996 A).

The definitive diagnosis of female genital schistosomiasis requires an invasive technique (a cervical biopsy) which is not part of a standard gynecological examination. And as the signs and symptoms of FGS mimic those of sexually transmitted diseases, unless physicians are especially alert to the possibility of FGS they are likely to overlook it. As reproductive health problems, especially those associated with sexually transmitted diseases, are surrounded by secrecy and silence, women with such problems rarely visit a doctor (Poggensee et al. 1999: Feldmeier et al. 1995).

Gender issues in diagnosis

Non-parasitological diagnostic procedures

So far in our discussion of screening procedures we may have given the impression that there was no alternative to parasitological investigation of urine and stool specimens for the diagnosis of schistosomiasis. But in fact, several alternative procedures have been developed. The question we want to pose here is to what extent are these techniques gender-sensitive, or indeed culturally sensitive?

A dipstick or reagent strip that, within seconds of immersion in urine, can indicate the presence of blood, could be a useful diagnostic tool for identifying *S. haematobium* infection. However, from the point of view of gender sensitivity, this technique leaves much to be desired, for the presence of menstrual blood in the urine would result in a false positive test result. Because of this problem, one study of school children in Minya governorate specifically omitted girls, examining only fifth grade boys (Sadek et al. no date). A recent Egyptian field study found that reagent strips had only a limited value under Egyptian conditions, especially for women and girls (Hammad et al. 1997; see also Feldmeier et al. 1993; Gunderson et al. 1996).

Researchers at the Schistosomiasis Research Project developed a single dipstick for blood samples (rather than urine) to identify antibodies for both *S. mansoni* and *S. haematobium*. They recommended it for screening in Upper Egypt and in newly settled areas, where prevalence was considered to be too low for the mass treatment (without prior diagnosis) that was, by the late 1990s, becoming an important aspect of MOH control strategies (El Khoby et al. 1998). This technique requires taking a blood sample from villagers, which presents different issues than does taking a urine or fecal specimen. Researchers taking blood samples in the Delta and Upper Egypt for a study of hepatitis C virus found that villagers considered urine and feces as waste substances which would normally be lost anyway, while they regarded blood as a non-replaceable sutbstance. Thus, they were often unwilling to provide blood samples or allow samples to be taken from their children.

Ultrasonography
Ultrasonography is an established method of identifying soft tissue damage possibly associated with schistosomiasis infection. Researchers and clinicians usually regard this technique as noninvasive as the patient is not expected to provide bodily fluids or to submit to an internal examination. However, researchers outside Egypt who have used the portable machine in a community setting have pointed out that women do not share this view. In an ultrasonography examination, the abdomen has to be uncovered, and in the case of *S. haematobium*, also the pubic area (Feldmeier et al. 1993). If the attending technician or physician is male, many women are likely to refuse to be examined. Another limitation of ultrasound is that it should not be used for pregnant women, as it may be difficult to distinguish between changes due to schistosomiasis and those due to pregnancy (Richter et al. 1996 B).

Reflecting on the use of the portable ultrasound machines in the EPI 1, 2, 3 study, one investigator said that women were concerned about inadequate provisions for a private examination, and did not like being examined by male doctors. He also thought that a larger number of men than women were examined as men were often more aggressive in seeking ultrasonography. However, in the published analysis of these studies no gender analysis was presented (El Khoby et al. 2000 B). It is clear that more information is needed about the extent to which women's experience of ultrasonography may be different from that of men, and what can be done to correct any imbalance in the number of males and females who may undergo ultrasonography.

Treatment with praziquantel—a gender-based assessment

As we have seen, praziquantel was established as the national standard treatment for *S. mansoni* and *S. haematobium* by a directive of the Ministry of Health in 1988. Yet since then, no official information has been available about the proportion of women, compared to men, who have been treated or cured. This was an important issue, as women who were pregnant could not be treated.

When praziquantel was undergoing its final field tests, it was shown to be safe for women who were taking oral contraceptives; this news was welcome in Egypt in the light of the widespread official support for family planning programs. However, no standard trials were conducted that included pregnant or lactating women. Because of the Ministry's fear of jeopardizing a large-scale drug treatment program, in Egypt the drug is not administered to pregnant women.

In 1995, the Egypt Demographic and Health Survey found that 10% of women of reproductive age were pregnant at any one time (personal communication, Dr. Eltigani E. Eltigani). This proportion would be higher among young women in their twenties and early thirties who often have a relatively high rate of schistosomiasis infection compared to older women. Unfortunately, we have no information from the MOH and only fragmentary data from epidemiological studies about how many women were actually not treated because they were pregnant. A study of the efficacy of mass treatment in a small village in Giza governorate gives a rare insight into the impact of pregnancy on treatment. The survey of all villagers identified 40 pregnant women (4.1% of the total population); 32 were tested and 5 (15.6%) were found positive, compared to an overall community prevalence rate of 23% (Talaat and Miller 1998).

Looking at our own epidemiological surveys, our records provide some indications that women were not tested or treated because they reported that they were pregnant. In the first survey, in 1992, of 48 females found infected in al-Salamuniya seven were not treated, mainly because they were pregnant.

Until the risks, if any, of the treatment of pregnant women with praziquantel are assessed, health providers could manage anemia due to schistosomiasis with iron supplements and folic acid (Kusel and Hagan 1999). However, the failure to follow up such women and provide them with treatment could leave a reservoir of infection in a community. This suggests that there is a need to target young women for treatment and prevention, even though they are not a group usually considered as being at risk. The need to target this group was also pointed out in chapter 9, when we studied the patterns of exposure to canals during women's domestic activity. In the next chapter we will discuss community-based approaches to schistosomiasis control that can involve women, as well as men and children.

12

Community-based Approaches to Schistosomiasis Control

Prevention: building on local awareness, skills, and needs

This chapter is largely devoted to a discussion of health promotion and other community-based approaches to schistosomiasis control. These approaches are designed to enhance the basis of local knowledge, foster sustainable behavior, and bring about environmental improvements that are crucial for the *prevention* of schistosomiasis transmission.

Essential elements of these approaches are, on the one hand, local people's own awareness and concern about schistosomiasis, and on the other, the skills they possess to meet the community's needs. Women and men have different roles in meeting these needs, as each brings particular skills and knowledge to the resolution of specific problems. Another element essential to schistosomiasis prevention strategies is recognition that they must relate to the lived experience of local people. If they do not, the strategies will fail. This is the most basic rationale for fostering local involvement (we prefer to use this phrase, rather than the overworked and idealized term "community participation").

During our period of study, from 1991 to 1996, the residents of al-Garda and al-Salamuniya knew they could be treated for schistosomiasis at the local health center. But while they did take advantage of opportunities for treatment, they seemed to be less concerned about taking action to prevent infection. Although they recognized that its transmission was somehow associated with canals, as we have shown in chapter 9 many of them considered that they had good reason to continue to use the canals for agriculture, for household chores, and for recreation activities. Improving awareness of all sectors of the population—women, men, and children—about the role of canals in transmission provided a starting point for the development of health promotion activities, especially with schoolchildren; we will discuss this in the first part of this chapter. We will then explore a number of other ways in which community-based initiatives might contribute to other schistosomiasis control activities, concluding with the planning of a public laundry.

Health promotion in schistosomiasis control

Health promotion is a useful concept which brings together the various strands involved in schistosomiasis control in the community setting. The Ottawa Charter for Health Promotion described it in 1986 as "a process of enabling people to increase control over, and to improve their health" (Ottawa 2000). Building on the definition set forth in the WHO Alma Ata Declaration of 1979, the Ottawa Charter interpreted "health" as a positive quality of life, rather than as a negative state, "dis-ease." Following this line of thinking, health promotion is largely concerned with preventive strategies. In addition to local involvement, however, practicing medical doctors and anthropologists now recognize that such strategies must also involve a national government's recognition of the essential nature of its own role in securing funding for the large scale infrastructure improvements (such as upgrading water, sanitation, and canal systems) without which community-level health education can do little more than effect cosmetic change.

For us, as researchers at the local level, an important aspect of health promotion is its essentially participatory orientation, rooted firmly in the locality, as something that is done *with* local people, not *on* them, *to* them or *for* them. It is a process, an activity directed toward enabling people to take action. Thus, it is the *process* itself that is the center of attention and of evaluation (Nutbeam 1998).

We will begin by exploring some ways that health staff interact with local people beyond the confines of the health unit, and how other local government employees such as school teachers, as well as formal and informal community leaders, can collaborate with residents in participatory health promotion activities. We will first examine existing arrangements for health promotion at the rural health units and through the mass media.

The role of the Rural Health Unit in local health promotion

Ever since its founding in 1977, the Egyptian National Schistosomiasis Control Program has stressed that health education is an important part of its integrated program. However, the staff of the rural health units (who were responsible for actually providing screening and treatment services for the general public and school children) were not assigned a role in health promotion. This meant that, as of the early-mid 1990s, there were no continuous face-to-face health education programs delivered directly by local health staff. Instead, health education was provided by a district-level team in the form of periodic "campaigns" organized around specific topics, including, occasionally, schistosomiasis.

The administration of these health campaigns was the responsibility of the Health Education Department in the Ministry of Health, which prepared topics and materials; from there they were sent down to the Department of Health in the *markaz* (district). District level health education specialists then went out to the villages in their "mobile units," equipped with their own materials and schedules. The district level team did not request, or expect, any cooperation from local RHU staff during the campaign. This approach was top-down, featuring instructions coming from outside, and directed toward individuals rather than toward specific sectors of the community (such as part-time farmers), or to families.

At the time we were doing our research, no staff at the health unit in al-Salamuniya or Baguriya, or the unit in the al-Garda mother village, were assigned to health education activities. For their part, RHU staff did not see this as one of their responsibilities (on this, see chapter 11). Furthering this segregation of activities, the Ministry made no provisions to train local staff on particular health topics unless they had received a grant for a specific project. Examples of training as part of a special program include family planning, diarrheal disease control, and maternal health projects, supported by USAID. No special programs have focused on health education for schistosomiasis.

Thus, it was not surprising that no materials on schistosomiasis were available at the al-Salamuniya Rural Health Unit for the use of the staff or for public distribution, although there were posters on the walls relevant to current child and maternal health programs. However, tucked away in a back cupboard were some booklets, posters, and handouts that had been prepared by the Health Education Department in the 1960s specifically for rural audiences. Appropriate to the schistosomiasis situation at that time (but not to the 1990s when *S. mansoni* had become the dominant form of schistosomiasis locally) they focused on *S. haematobium*. They have never been updated.

The reason why there was no relevant printed material on schistosomiasis in rural health centers was because the role of health education had been entirely taken over by mass media messages on the TV and radio. It seems that in 1988, at the time when praziquantel was officially adopted for general use, the Ministry of Health had allowed itself to be persuaded that the most effective way to communicate with the public was through radio and the new medium, TV. In its attempt to get its message across it employed experts in "social marketing" who tried to sell "health" with the same techniques used to sell soap and other household cleaners.

Mass media health messages

Beginning in the late 1980s, a number of TV messages about schistosomiasis were shown repeatedly throughout the country. Most consisted of one minute spots, inserted between programs, and others were conveyed during soap operas. Similar techniques were used on the radio. Mass media strategies of this kind had the potential to raise awareness among a large audience; they were a critical first step in health promotion. As we saw in chapter 7, 90% of households in al-Garda and 83% in al-Salamuniya had TV. In both communities the vast majority of householders told interviewers that they received their information about schistosomiasis primarily from TV. But this raises the question, what was the quality of the information they came away with?

While some messages appeared to be effective, such as those that encouraged people to go to rural health centers for free diagnosis and treatment, others had less impact. Sandra Lane (1997) pointed out that many of these slick messages were not adequately pre-tested or evaluated to ensure that they conveyed their message effectively. Loza also pointed out that there was little evidence that they had led people to change their behavior at the canals (Loza 1993: 39).

In the early 1990s, the TV message that residents in al-Garda and al-Salamuniya mentioned most frequently was: "Keep your back to the canal." This was similar to the message that had been widely used in earlier health education programs, so it was not new to audiences. However, our exploration of behavior at the canals showed that this advice was neither practical nor feasible for farmers and many other local people (see chapter 9). Indeed, members of the research team observed young children cheerfully singing the TV ditty about not going to the canal, while splashing about in its shallow waters. The TV message had obviously failed to change local people's behavior.

Planning health promotion to change behavior

Health educators have come to recognize the limitations of the "exposure-to-messages approach," such as the TV spots that we have just mentioned (Quarles 1994). They recognize that the relationship between a person's knowledge of a health risk and that person's actual behavior is problematic.

In Egypt, many cigarette smokers have been alerted to the fact that smoking causes cancer, yet they continue to smoke. Similarly, many rural men, women, and children know that entering the canal puts them at risk for schistosomiasis, yet they continue to do so. The real measure of the impact of a health education program is a change in behavior, and not just a change in knowledge or attitudes. Moreover, the change in behavior needs to be sustained, to become a routine part of daily life.

Working in al-Garda and al-Salamuniya, our research team enlarged upon a model we had formulated in an earlier research project on environmental and hygiene education in rural areas of Munufiya governorate. This earlier study had also been participatory, and, like the schistosomiasis study, involved local people and used existing resources and organizations (El Katsha and Watts 1994A). Table 12.1 lists the stages in the health promotion model we developed for local activities related to schistosomiasis. These involved local staff and local community members as partners in developing an appropriate and sustainable program.

Table 12.1: Schistosomiasis health promotion activities

1. Problem identification: constraints and resources
   Activities at the canals, and residents' explanation for them; contamination behavior;
   Barriers to effective diagnosis and treatment at the health unit
   Identification of resources: knowledge and skills of health staff, teachers, adult residents and schoolchildren.
2. Identification of target audience(s)
3. Training of educators—including communication skills
4. Design and pre-testing of materials
5. Education on schistosomiasis
6. Monitoring
7. Evaluation—feedback and modification, if needed

Supporting activities:

1. Upgrading local diagnostic and treatment services
2. Working with local people to improve the environment—water, drainage/sanitation, canals.

In the course of our first stage, problem identification, we sorted out which issues lent themselves to the "health promotion" strategy at the local level, and which issues could only be resolved by government agencies. In the second category, for example, was the need to establish and maintain a safe drainage/sewerage system. Part of the "problem identification" stage of our project was to pinpoint the areas that health educators could do something about at the local level.

The health promotion activities described in this chapter consisted of face-to-face interaction in an informal non-didactic setting; formal lectures were never used. The researchers and their partners targeted specific groups, such as women attending the RHU and older primary school children reached by the school screening program. As much as possible, we entered into ordinary people's lived world of everyday experience. As such, our approach was very different from that used by the mass media which usually targeted an undifferentiated audience "out there." Designed as an ongoing, sustainable local activity, our approach was also different from the periodic local "campaigns" that characterized the activities of the Department of Health Education at the time of our research.

In designing this approach we were aware of the large number of theoretical frameworks for health education that were directed toward behavior change. Helmut Kloos, for example, in a paper specifically directed toward schistosomiasis, suggested a PRECEDE model as a diagnostic tool for health education planning, intervention, and evaluation. This model classifies various "predisposing," "enabling," and "reinforcing" factors that contribute to improvements in health or a decrease in disease levels. Kloos (1995) and others (Robert et al. 1989) stress that in designing health education programs it is essential to use local information that is directly relevant to the area where the material will be used. Such information includes local people's knowledge of what promotes good health, their health behavior, and their rationale for such behavior. This insight about the subject largely coincides with our own ideas.

Health promotion in schools

We have already identified (in chapter 3) the many reasons to target school-age children for schistosomiasis screening, treatment, and health education programs. These include their high rates of infection and reinfection, and their high egg loads, compared to other segments of the population. Moreover, those who attend school are a "captive audience." In al-Garda and al-Salamuniya, around 90% of children aged 6 to 14, girls as well as boys, regularly attend school. Children of this age have long been known to be especially responsive to health promotion programs. They are willing to learn, responsive to creative educational methods, and often communicate information to their peers and to their parents. They are future citizens; in due course most of them will become parents (Bundy and Guyatt 1996; Morley 1993).

For many years the Egyptian government and NGOs have been supporting programs directed toward improving the health and health knowledge of school-age children. More recently, the World Health Organization created a Global School Health Initiative. This encourages innovative research into health promotion in schools, and attempts to strengthen institutional support networks for such programs (WHO 1998). Here, we will focus on what actually happens at the local level, as well as suggesting what can happen with appropriate support from the community and from school staff.

Our health promotion activities focused on primary school children between the ages of six and ten or eleven. As most Egyptian communities of any size have at least one primary school with a large proportion of children attending, the strategies we developed and tested in al-Salamuniya and al-Garda can be further tested, and replicated in similar settings elsewhere in Egypt.

Our initial research in the two communities revealed that teachers themselves had a limited knowledge of schistosomiasis, especially about *S. mansoni*, the disease form which in recent years had become dominant in the communities in which they lived and taught. There were several reasons for this. Older teachers had learned something about *S.haematobium* (detected in urine) during their time in teachers training college. However, younger teachers knew little about the disease, as the curriculum of the new general university course for primary school teachers, that was replacing the teachers training college course, did not include a unit on schistosomiasis. Therefore, the researchers agreed to prepare a booklet on schistosomiasis that teachers and students could use.

In 1988, the sixth grade teaching module on schistosomiasis was dropped from the primary curriculum when the final year of primary school was abolished. Thus, at the time of our research, in the early 1990s, one of the first challenges was for teachers to identify other opportunities in the crowded curriculum in which they could introduce this topic.

Our first task was to identify the teachers and other school staff, such as the school nurse, who would be willing to participate in a health education activity. We then conducted interactive workshops for these volunteers in which we encouraged them to develop ways to interact with students, rather than simply subjecting them to lectures. We encouraged them to give positive messages encouraging safe behavior, rather than negative messages forbidding certain activities.

Researchers, teachers, and the school nurses also met together to discuss when and how material on schistosomiasis could be presented to students, and teachers were encouraged to try out these ideas. Two Arabic language teachers suggested that structured, interactive activities could take place in the weekly "listening and learning" class and volunteered to introduce a special project on schistosomiasis. They mentioned as a precedent the students in one al-Salamuniya class who had made a book about their experiences of the 1992 earthquake. Other teachers used schistosomiasis as a topic for composition, or had students make posters in art classes.

Other activities were undertaken outside actual class time. For example, in two schools, the school nurses gave a short message about schistosomiasis in the regular assembly held at the beginning of the school day. Moreover, in all participating schools copies of the booklet prepared by the research team and a group of teachers were put in the school library. In one school, the volunteers in the school-based Red Crescent program (mostly girls) enthusiastically participated in a schistosomiasis program, posting information on school notice boards and talking to their fellow students about it.

After the al-Salamuniya teachers had used the booklets for three months, they individually told researchers about their own and the children's reactions. Taking these comments into account, we revised the booklet. Then we had teachers in al-Garda test the revised edition with 800 third, fourth, and fifth grade primary students. This, too, was a useful exercise. The al-Garda teachers suggested that the pictures needed to be in color if they were to fully engage the children's attention and that the diagram of the transmission cycle needed to be simplified. We incorporated these suggestions into the revised booklet, and a copy was submitted as part of our final report to the Schistosomiasis Research Project.

Although most people in Egypt had heard on TV about "microbes" (in Arabic, *mikrubat*) and other tiny living disease-causing agents, we found that the al-Salamuniya primary school children found it difficult to accept that invisible creatures and tiny snails could be responsible for the transmission of schistosomiasis. To help them overcome this difficulty, the local snail inspector arranged to take several groups of older primary school children for a walk along the canal. He explained what kind of snails he was looking for, caught some in a net, crushed them and placed them under the microscope. The children were thrilled to actually be able to see the schistosomes that caused them so much discomfort. At the same time, the snail inspector told them which sites in the village were most likely to infect them. This excursion with the snail inspector was such a success that we explored the possibility of establishing such an activity on a regular basis. However, we found that this would require collaboration with the staff of the Ministry of Education beyond the local level, and could not be done solely on the initiative of local teachers.

Our qualitative evaluation of the processes of health promotion among primary school children in al-Garda and al-Salamuniya showed that teachers could play an important part in developing and testing schistosomiasis health materials as well as being effective health educators. They are authority figures to whom most young children listen with respect. Moreover, in our two study communities, as in most other places in rural Egypt, primary school teachers are local people with close ties to the community, and are well known to parents and children outside the school setting. Within the school, they are operating in a role that is comfortable and familiar to them, and to the children. Our study showed both that teachers can be creative and innovative when they are given encouragement and appropriate educational materials, and that children will respond positively when they see that their lessons in school relate closely to their daily life.

Summer clubs

At the same time as we were exploring what could be done *in* school, we were also looking at alternative, out of school activities. Some of us had earlier developed and tested a program of summer clubs in other communities in Munufiya governorate. These focused on hygiene education and environmental health. The summer club program had begun in 1988 in two villages and then extended, by the third year, to cover ten villages. By the mid-1990s, the program, which by then included a component on schistosomiasis, had reached over one thousand children.

The summer clubs, held in the four-month long summer vacation while the children were at home with their families, provided an ideal opportunity for teaching life skills and linking what was learned in school with out of school activities and behaviors. In these settings, teachers were not bound by the demands of a fixed curriculum that largely relied on rote learning. The children attending these summer clubs were all fourth and fifth graders, aged between nine and eleven. At that time, the fifth year was the final year of primary school and the last uninterrupted year of schooling for some children. As the summer clubs lasted for one month, and met three times a week for a three-hour session, there was still plenty of time for girls to help their mothers with housework, and for both boys and girls to work in the fields.

The summer clubs were a spin-off from a health education program one of the authors had developed in association with water and sanitation interventions in two villages. In pre-intervention studies (in 1987) both teachers and children identified serious local environmental problems. The children were aware of the health risks of using canals for swimming and domestic activities, and knew that people should not throw garbage and wastewater into these waterways. One module on environmental sanitation related to these concerns; other topics included personal hygiene, nutrition, and food handling. Within this framework, activities were relatively unstructured, with the children doing group work, making posters and craft items, story telling, and acting. The teachers also took the children on a walk around the village to identify environmental hazards. The children then took part in village clean-up campaigns, collecting rubbish and sweeping the alleys and the areas in front of their family houses. The researchers prepared a brief teaching guide on health and the environment, which was tested and used by the teachers (El Katsha and Watts 1994A; 1994B; 1993). The introduction of a module on schistosomiasis, using the booklet already prepared by the research team in al-Garda and al-Salamuniya, fitted in well with this summer club program.

In 1997, the summer school program was introduced to children in six schools in two areas of Ismailia governorate, covering hygiene, environmental health, and disease transmission, including schistosomiasis. It also included a range of activities such as crafts, small group activities, a village walk, and a clean-up. It reached 577 students. Pre- and post-intervention surveys showed a significant improvement in the children's knowledge of the causes and health impact of schistosomiasis (Mehanna et al. 1998).

Social capital

As we have pointed out a number of times in this book, the effectiveness of strategies for the control and prevention of schistosomiasis depends on the degree of cooperation and mutual trust between government, on the one hand, and the rural people who live in a schistosomiasis-threatened environment, on the other. Many aspects of a successful control program (or, in the long-term, a program for the total eradication of the disease) ultimately depend on firm action by government.

It is not to be expected that individuals or networks of families living in rural areas will have the financial capital and political clout required to build safe sewage disposal systems for settlements with ten thousand inhabitants. Similarly, local voluntary groups and NGOs cannot establish a regular effective system of sanitary inspections and enforce regulations against the dumping of sewage into irrigation canals unless these activities are supported by local government.

No ordinary individual or group of families living in rural areas in the first years of the twenty-first century can be expected to have the capability to reassess the suitability of the nineteenth century irrigation system that still forms the basis of Egyptian networks. Certainly, no private individual of ordinary means has the capability to draw up plans for a properly engineered, disease-free irrigation system.

In the early years of the twenty-first century, in many parts of the world, many major infrastructural projects are being undertaken by for-profit multinational companies. However, responsibility for the control of the behavior of multinational organizations still rests with the governments and law courts of individual nation states. Thus, it is up to central government authorities to decide which (if any) aspects of a long-term schistosomiasis control or eradication program (necessarily involving safe water and sanitation and a re-thinking of the irrigation system) might be entrusted to non-governmental, for-profit, multinational organizations and which aspects of the program must still be undertaken, in unified and coherent fashion, by the central government itself.

At the other end of the equation is the resource known as "social capital." As defined by R. Putnam in 1993, "social capital" consists of self-help networks of local people, associations and partnerships that exist in a given social setting and that are able to involve people in various activities (Gillies 1998). Casual observation, combined with the assumption that Egyptians believe that they should leave all such initiatives to "the government," has often lead commentators to characterize the Egyptian village as an unfavorable setting for cooperative activities, with people acting on an individualistic basis, rather than through groups and alliances.

As we have seen, a rural "community" in Egypt can rarely be described as homogeneous, acting together, in the long term, in pursuit of common interests. However, people in a small village or in a neighborhood of a larger village can unite (often on a short-term basis) to undertake various forms of collective action (see, for example Saad 1998). A recent study showed that local residents in relatively poor neighborhoods in Cairo were acutely aware of pollution in their immediate environment, and had established local groups willing to take action. In the highly polluted settlement of Kafr al-Elow, near Helwan (a heavy industry area south of Cairo) residents collaborated in garbage collection and in removing material dredged from the canal that was thrown in foul smelling heaps along the canal-side roads and paths (Hopkins et al. 2001: 135–39). Only time will tell if this initiative results in a strong local organization or if residents succeed in getting government support for more permanent solutions to their problem.

We have earlier noted concrete examples of cooperation between neighbors in al-Garda and al-Salamuniya during our research period. For example, in chapter 8, we saw that, in al-Salamuniya, men from households along the same alley (following their expected roles in dealing with certain "authorities" in the village) got together to negotiate with local staff, and to share the cost of constructing a feeder pipe to link their houses to the sewerage system. In another village in Munufiya governorate, women (responsible for the use and disposal of domestic water) collaborated to maintain a shared standpipe, and their husbands lobbied the local authorities for repairs (El Katsha and Watts 1993). Most large villages have a Community Development Association that develops programs in response to local needs. It is supported by the Social Unit and registered with the Ministry of Social Affairs as an NGO—as such it occupies a somewhat anomalous position between a government and a non-government organization.

Community-based health promotion for schistosomiasis control

During our research in al-Garda and al-Salamuniya we found that many residents recognized that their communities suffered environmental problems, which they associated with the pollution of canals, unsafe disposal of septic tank effluent, wastewater, and garbage, and that these forms of environmental degradation adversely affected their health. They also knew that the canals were the source (or for some of them only one of several possible sources) of schistosomiasis.

They certainly recognized that some sort of balance necessarily existed between what Government should and could do to reverse locally-present environmental degradation processes and what they, as community residents and citizens, could do for themselves. As might be expected, there was considerable variety of opinion as to what the precise nature of this balance should be. Some people laid the onus for action almost solely on the Government; given Egypt's long history of centralized rule, this view is not unexpected. However, a substantial number of villagers felt able to tell our research team that "the solution is up to the people themselves." As one respondent stated: "The people are responsible for the cleanliness of the village." Provided with an enabling framework, many rural people in communities in Munufiya have shown themselves willing to contribute to such programs, in spite of their busy schedules.

In al-Garda and al-Salamuniya, toward the end of our project, the research team and local people who had participated in it met with other residents who had so far not been involved, or only peripherally involved. The objective of these meetings was to share what had been learned about schistosomiasis and to discuss what could be done to further limit transmission. The researchers also wanted to know if the residents agreed with their general approach and with their findings. These meetings attracted a large audience of men and women and generated lively discussions.

In conducting these meetings, the research team focused on al-Salamuniya, as it had a much higher prevalence of schistosomiasis and, unlike al-Garda, it had both a Rural Health Unit and a Community Development Association. Members of the governing board of the CDA, who included the doctor at the RHU, requested that we provide them with a report, in Arabic, of our main findings; this we did. Our report was then presented before an open meeting held in the CDA meeting room.

It was at this meeting that the primary school headmaster, in presenting the findings, referred to the project as "our research," indicating that he and other local people felt a sense of "ownership" of the findings which they had been instrumental in uncovering. Seen in this light, our research team had served only as a catalyst in helping local people to formulate ideas and concepts with which they had already been familiar.

The CDA audience was particularly interested in hearing about the symptoms of *S. mansoni*, about how they could protect themselves from infection, and about the environmental conditions that encouraged disease transmission. The researchers identified a number of canal sites in the village that, at that time, were most likely to transmit infection, and others that were less likely to do so. Another meeting between CDA members, researchers, and other interested people was held to discuss further community involvement in schistosomiasis control. On this occasion the CDA expressed its interest in pursuing the feasibility of a drainage/sewerage system. They, and we, knew that their expression of interest was a vital first step but that any further progress on a project of this sort would necessarily have to involve high-level government officials at the district and governorate level.

A proposed laundry intervention

Planning for the laundry intervention

In dealing with schistosomiasis endemic areas around the world, planners have considered public laundries as a way of providing facilities for local women who would otherwise have used rivers or canals and thus been at risk of schistosomiasis. Communal laundries played such a role in a successful project for *S. mansoni* control in Saint Lucia, alongside individual household water connections and treatment with drugs (Jordan et al. 1982).

In large villages in Egypt, the establishment of communal laundries is, by its very nature, beyond the immediate capability of local residents. Facilities of this sort obviously require secure control of a plot of land (which is very expensive), as well as a safe and reliable way to supply water for washing and then to remove the wastewater. In 1994, the Schistosomiasis Research Project asked our team to monitor a proposed public laundry project in a community of 3,500 people in Munufiya governorate that we will call Baguriya. The laundry was planned specifically to limit schistosomiasis transmission.

Over a period of eighteen months, our research team monitored the planning phase of this laundry. Our objective was to involve local women (and men) in the design and siting of the facility, in order to ensure that it met their needs and would be *used.* If it was not used and women continued to wash in the canals, it would not have any impact on local schistosomiasis infection rates.

In 1994, our epidemiological survey revealed that 11% (5 of 45) of Baguriya women between the ages of 15 and 34 were infected (our limited funding only allowed us to take a 10% sample of the population). Our water contact studies in the village revealed that Baguriya women's behavior was similar to that of women in al-Garda and al-Salamuniya. As we have seen in chapter 9, many young women in these two communities regularly used the canals, and in al-Salamuniya one third of those aged between 15 and 34 were infected. Thus, one could reasonably expect that a functioning public laundry, *if used by local women*, would have a positive impact on infection.

The proposed laundry in Baguriya was a pilot project paid for by NOPWASD (National Organization for Potable Water and Sanitary Drainage), part of the Ministry of Housing and Reconstruction. Once established, it was to be administered by the Social Unit in Baguriya. Technical assistance was provided by a bilateral aid agency, and the initial plans were prepared by an engineer (a European woman) from the company that was to implement the project. We collaborated with her while we were working in the village.

The Village Council had already earmarked a plot next to the main canal as the site for the laundry. However, they appear not to have done their homework. Further investigation revealed that the land actually belonged to the Railway Authority and this government agency was not prepared to hand over the land for a laundry. This was the reason why the facility could not go forward during the period for which we were funded. If nothing else, this focused our attention on institutional barriers to effective disease control interventions in Egypt.

Our first task as researchers (who of course did not know in advance that the project would not be carried out) was to find out what Baguriya women expected of a communal laundry. Most wanted a laundry that replicated the practices with which they were already familiar at the canal. They wanted to wash in "sweet" (i.e. canal) water, and have a drainage system to get rid of wastewater. Some women wanted to wash dishes as well as clothes at the laundry, and asked for a separate washing area and garbage containers for food waste and other rubbish. Some also suggested that a space be set aside for pre-school children to play, so that they could watch them while they were washing.

Most women in Baguriya said they had known about the link between canal water and schistosomiasis for many years. However, in our survey of 180 women, only one third of those who said they were interested in using the proposed laundry (41 of 115) stated that they were motivated to do so by fear of schistosomiasis. In terms of the risk posed on a day to day basis, women considered that schistosomiasis avoidance was a low priority. Perhaps this was because they calculated that if they were infected they could simply go off to the local rural health unit and be cured, free of charge.

Baguriya women stated that they would be willing to assist in keeping the laundry clean but had no clear idea how this could be arranged. Someone suggested that this task be assigned to a responsible employee of the Village Council. However, when a group of women from Baguriya visited a public laundry in a nearby village, and saw that it was being supervised by a male employee of the Village Council, they were most emphatic that this was simply not acceptable—for a male attendant should not intrude in a woman's space.

On the question of payment for the use of the laundry, some women agreed that they should pay, on the grounds that people do not value what is given to them free. However, others questioned why they should pay for the laundry while canal water was free.

The original plan for the laundry, prepared by the aid agency, included fully automatic washing machines. One woman expressed her delight at the prospect of using such machines, and indulged in some wishful thinking: "We will send our dirty clothes and dishes and pick them up later, cleaned, washed and dried." However, in their discussions with us, the majority of women were concerned about the constraints imposed by automatic machines, which required fixed, predetermined blocks of time, imposed by the washing cycles. They were worried about the need to take turns using the automatic machines. By way of contrast, there was always plenty of room at the canal side for any woman who wished to wash any time. They also expressed unwillingness to leave their clothes in the laundry and do domestic tasks elsewhere, yet they felt they could not simply sit around and talk to whoever was waiting for their washing cycle to finish. In general, the Baguriya women recognized that the whole spatial, social, and temporal structure of their domestic world would change if they opted to use an automatic laundry at a communal site. However, they expressed their willingness to try out the automatic machines.

Guidelines for local involvement in a laundry intervention
In the hope that we will encourage others in planning and health to involve community members in project planning, we set forth our strategies for local involvement in table 12.2.

Table 12.2: Stages in local involvement for a laundry facility

1. **Planning the facility**
   Familiarization by planning authority; including initial meeting with village leaders, village meeting(s).
   The initial decision that a facility is needed, or not
   Needs assessment—patterns of water use, water supply, waste water disposal etc. (if possible done by local residents)
   The design of the facility
   Planning for use—maintenance, supervision, payment, opening times etc.

2. **Mobilization to use the facility**
   Involving women and men

3. **Maintenance of the laundry**
   Who will do it on a regular basis—paid supervisor, voluntary helpers
   Monitoring of site and facilities

Ideally, consultant planners, and the village people for whom an intervention (such as a laundry) is designed, should agree in advance that it is needed. Indeed, the initiative for the intervention should come from local people. But, in the real world, planners rarely look to local people for such initiatives. And, when an outside agency singles out their community for an intervention, local people see it as a sign of prestige for the whole community, and are unlikely to turn it down.

We discussed aspects of mobilization and maintenance with Baguriya women. Even at the planning stage, they were emphatic that men, too, should be involved. They explained that their husbands needed to understand what was going on and what the laundry would be like. As they pointed out, men had the power to forbid their wives to use the laundry.

Enabling local involvement
In this pilot project in Baguriya, the researchers were specifically invited to help the planners involve local people in designing and maintaining a public

laundry. Participation requires flexible guidelines so that, for example, local people can make suggestions about the design of the laundry and about how the service should be paid for and maintained. Local people also need to have a reasonable expectation that their views will be listened to. In terms of planning, there should be scope for all participating parties to explore a range of possible laundry solutions. For example, in a 1995–96 study for proposed laundries in three hamlets in the northern Delta, women identified a low tech option as appropriate, as the settlements were far from a Village Council that could be expected to provide maintenance staff. In this case, the women were quite prepared to work together to ensure that the laundry, a simple private shelter with low washing areas, would be kept clean.

As of 1994, when the laundry in Baguriya was being planned, the national agency NOPWASD had established no guidelines for the participation of villagers in planning activities. Like top-down health education, this was an operation planned outside the community and usually delivered without any consideration for specific, local needs. NOPWASD practice was not to charge for the use of the laundries at first, to encourage women to use them, and to introduce charges later, when people had become familiar with them. As villagers pointed out, how can one expect to pay for something that has once been free, especially given that access to the banks of the canal was still free and unrestricted.

As we have shown in our discussion of the Baguriya laundry, so long as there is some degree of trust between local people and outside planners, women (and men) are willing and able to contribute to the planning process. We believe that such local involvement is the best way to ensure that the laundries will be accepted and actually used. In a case of this sort, local women canal users who see a canal setting as the best place to wash their laundry and their dishes will be persuaded to use an alternative, schistosomiasis-free facility. A holistic approach to the resolutions of situations currently hazardous to health thus necessarily involves people at all levels, from government authorities at the center and at the governorate and district levels, to planning consultants and health specialists, and partners at the village level.

13

Conclusion

Understanding the complexity of schistosomiasis control

Our in-depth study has explored the many facets of schistosomiasis control in two rural Delta communities, examining both the human situation and the environmental setting. In presenting our findings we have looked at gendered human behavior as a way of providing access to the complex of local understandings of the illness, its transmission, diagnosis, treatment, and possible strategies for prevention. We cannot identify a universal panacea for schistosomiasis control for the new millennium. However, we hope that, in presenting a view from the village, as seen by villagers and by health providers at local health facilities, we can suggest some ways to move forward.

Our bottom-up approach is rather different from that of a conventional epidemiological survey that presents findings based primarily on a statistical analysis of infection in a particular population. Nevertheless, the starting point for our study, which helped us to formulate the questions we initially asked, was our 1992 epidemiological survey. This indicated that the infection rate in the community with a lower rate of household water connections and without a sewerage system (al-Salamuniya), was three times that of the village with a higher rate of household connections and a sewerage system (al-Garda). However, we found that the epidemiological data did not actually demonstrate a clear statistical correlation between a household's lack of an individual water connection or sewerage and the likelihood that household members would be infected with schistosomiasis. These findings emphasized the importance of our holistic approach. We needed to learn more about the use and maintenance of water and sanitation facilities, about people's contamination and exposure behavior, and about the ecology of local canals within the built up areas of the villages and out in the fields.

As rural people are infected and treated in the rural setting (usually within a few kilometers of their home), we explored the ways in which the government primary health facilities delivered their diagnostic and treatment services, and the expectations of both patients and providers about these services. The "action" part of our research involved working with providers to improve strategies to deliver these services. We also explored the feasibility of a number of environmental interventions that could help to control the spread of the infection. This could be done either directly, by changing conditions in village canals, or indirectly by developing strategies to change the behavior of villagers that exposed them to the risk of infection at canals.

Gendered human behavior

As social scientists, our understanding of the groups we studied was informed by the recognition that human behavior is gendered—that girls and boys learn appropriate gender roles, attitudes and expectations as they grow up and then act them out in their daily life as mature adults. Our inquiries were designed to understand such behavior and to identify possible strategies for behavior change that would result in less transmission of the disease and more effective screening and treatment.

Our studies of exposure behavior at canals within the built-up areas of the village showed that groups of males and females occupied separate sites on the canal banks and carried out gender distinct tasks. In contrast, in the fields surrounding the village, family groups worked together irrigating crops and in other farming activities. This is the only study we know of that has attempted a comprehensive view of all community members' canal-side activity within the built-up area of the village (mostly domestic activity by women) as well as in the surrounding fields.

Overall, we found that the diversity and range of canal side activities was similar in both Delta communities. It was only by looking at the different types of exposure activities that we could assess their relative importance for disease transmission. Putting our water contact studies together with the evidence of infection rates among the various occupational groups as well as from information garnered during in-depth discussions in the villages, we concluded that farming, especially by males, was the major "risk behavior" for schistosomiasis. It was vitally important here to include part-time farmers for many adults went to the fields for part of every day after the end of their regular workday, and many women who called themselves "housewives" actually farmed on a fairly regular basis. We also found that farmers—who in the nature of things depended on irrigation—could not identify any way in which they could realistically protect themselves from infection. This prompted us to explore possible options to change the ecology of canals so that they would be less hospitable to vector snails.

At the same time, we found that women had their own good reasons for continuing domestic activities in village canals. This was an important risk behavior for young women, especially in al-Salamuniya where one third of women aged between 15 and 35 were infected. As a result of monitoring the planning stages of a laundry in a third village in Munufiya governorate we were able to develop a protocol to encourage the participation of women in all stages of the laundry operation. By bringing local women into the process, we hoped to ensure that the laundry (if built) would actually be used and would contribute to a decrease in infection among women.

Gendered human behavior also affects individuals' understanding of schistosomiasis, and their dealings with the local health unit that provides diagnosis and treatment. Our study of the actual functioning of the school-based screening program in al-Salamuniya showed that the collection of stool samples was a major problem for children, especially for girls. In response to this, we worked with local RHU and district health staff to design and test a revised, gender sensitive strategy for collecting stool samples. The new strategy was intended to ensure that girls and boys had an equal opportunity to be screened, diagnosed and treated, and that their need for privacy would be respected. Our team also worked with teachers on health promotion activities for girls and boys, both within school hours and during summer clubs in the summer vacation.

Environmental interventions for schistosomiasis prevention also require an understanding of gendered human behavior. In the two study villages, we explored the roles and responsibilities of women and men with respect to domestic water use and disposal, disposal of latrine effluent and garbage, and the maintenance of canals. We then asked how these roles and responsibilities could be incorporated into plans for local involvement and how local government-employed health and other staff could respond to these needs.

Our exploration of the scientific knowledge about schistosomiasis in Egypt, and in other endemic areas, indicated that an understanding of gendered human behavior was not built into biomedical or epidemiological studies. For us, one of the reasons for creating an interdisciplinary research group was to enable all members to work together to share this understanding of gender, so that it became "second nature" to consider gender in all aspects of the research. The research team also worked with health providers to develop this awareness, in the hope that it would infuse their dealings with patients with a new sensitivity.

Why environmental control?

Schistosomiasis is a disease that is transmitted by human activities associated in one way or another with water. In chapter 8 we saw how environmental problems—especially those associated with domestic water supply and sanitation, and the pollution and characteristics of canals flowing through the built-up areas of villages and the fields—formed a backdrop favorable to disease transmission in al-Salamuniya, and to a lesser extent in al-Garda. Environmental control is essentially preventive, and requires collaboration between the MOH and various other government ministries concerned with irrigation, agriculture, and water supply.

Toward the end of our research project, we explored the feasibility of some environmental interventions in al-Salamuniya, all of which would require the collaboration of several different ministries. Thus, our first task was to ascertain, in the course of discussions with various ministry officials, which ministry was responsible for overseeing the various activities. We focused on those activities that were feasible utilizing existing staff, skills, and facilities, and therefore could be replicated elsewhere in rural Egypt.

Changing the water flow in canals

We found that differences in infection rates in al-Garda and al-Salamuniya could, in large part, be explained in terms of the contrasting irrigation regimes in the two communities. In both villages, farmers were the most highly infected group, and the behavior of full-time and part-time farmers that exposed them to infection was similar. As farmers on irrigated land could not realistically change their water contact activities, we therefore asked how the irrigation system in al-Salamuniya might be modified to minimize the risk of infection.

The Ministry of Agriculture and Land Reclamation and the Ministry of Public Works and Water Resources are jointly responsible for controlling water flow in the larger canals, and any modification requires collaboration between them. Accordingly, we met with officials from both ministries to discuss the possibility of changing the level of the canals in al-Salamuniya, which at that time flowed slowly and did not dry out between irrigation turns, creating conditions that were more favorable to snail survival than in al-Garda canals (see chapter 8). We had hoped that this could be done by regulating the opening of certain canal sluice gates. However, the officials stated that, under conditions existing in the mid-1990s, it would be difficult to regulate the flow of water in the al-Salamuniya canals as the whole system is interconnected. For example, the al-Salamuniya canal that flows in front of the RHU has two entrances, but water flow is regulated by a single sluice gate, at the entrance to the Diya al-Kom canal. Thus, canals running through the built-up area of the village and the al-Salamuniya *zimam* never have a chance to dry out completely between irrigation rounds, or even during the annual winter cleaning.

In our discussions about plans to modify the flow of al-Salamuniya canals, we identified a more general problem. As of the mid-1990s, the Ministry of Agriculture and Land Reclamation and the Ministry of Public Works and Water Resources favored a continuous year-round flow of water in *all* canals in order to maximize farmers' access to water for irrigation so that they could grow whatever crops they wished throughout the year. At the same time, from a technical point of view, it also became feasible to limit, or even abolish, the annual winter canal closure that had previously been needed to clean the canals. By the early 1990s, the canals were closed for only two or three weeks. As the annual winter cleaning was now done by mechanical dredgers rather than manually, the Ministry of Public Works and Water Resources considered that it was no longer necessary to close the canals even for the two or three week period.

The decision to encourage a regular flow of water in the canals was taken by the two ministries concerned with agriculture, irrigation, and water resources. Yet, when seen from the perspective of schistosomiasis control, their decision was likely to cause an increase in the overall density of snail vectors in the canals, the opportunity for schistosomes to come into contact with the snails, and a massive increase in the risk of schistosomiasis. However, these ministries were not aware of the health implications of this decision and thus did not consult with the Ministry of Health.

Covering canals

Covering canals and drains flowing through densely populated villages would prevent residents from using them for recreation and household tasks. In communities such as al-Garda and al-Salamuniya, the covering of canals would also hinder people from constructing pipes to convey sewage and wastewater from their houses into the canal and dumping rubbish in the canal. As of the mid-1990s, the Ministry of Public Works and Water Resources had a national plan to cover canals, especially those that flowed through built-up areas. Al-Salamuniya was included in the plan for Munufiya governorate, but during the time we were working there, nothing was done. As of the year 2000, some canals and irrigation drains are being covered in both study villages, including the highly polluted stretch of canal in front of the health unit. Once this program to cover canals really gets under way, it will contribute to the control of schistosomiasis and other water related diseases in rural areas.

Weed control

Our snail studies in the two study communities indicated that the presence of water hyacinth and other vegetation was associated with high densities of vector snails. Removal of vegetation, as well as the removal of rubbish, would also reduce the amount of time that farmers need to spend in the water clearing vegetation from the intake pipe of their diesel irrigation pumps.

We discussed with an engineering consultant from the High Institute of Public Health in Alexandria the feasibility of clearing vegetation from the canals in al-Salamuniya. We found that the Agricultural Cooperative in the mother village was responsible for clearing the feeder canals, while individual farmers were responsible for clearing the *misqa*, field canals, which provide water to irrigate their plots of land. We suggested that the Cooperative, which clearly has the responsibility for clearing these feeder canals, should encourage men and boys to volunteer to clear the canals manually, as it would benefit all of the farmers who drew their water from them.

We also suggested that staff of the Snail Control Section of the Schistosomiasis Control Program should inform the Cooperative staff where infective snails had been found, so that they could give priority to clearing vegetation from those areas. As of 1996, intersectoral cooperation of this sort was not in evidence. However, there are sufficient staff in most cooperatives and Village Councils who could, with some training, be encouraged to implement canal clearing activities on a larger scale than was being carried out at that time. In the long term, however, the services of biologists are required in order to find out how best to control (or better still, get rid of) the alien weed, the water hyacinth, which is a cause of much of the difficulty.

Snail control

As of the early 1990s, under the government Schistosomiasis Control Program, chemical mollusciciding was the principal strategy for the control of snail populations. By 1988, in response to growing concern about the environmental impact of copper sulfate, the control program began to replace this chemical with the petroleum-based niclosamide, and to use it in smaller and more focal applications. However, in addition to chemical treatment and canal clearing there are other ways to target snails. One of these is the use of a local plant to kill snails.

A Ministry of Agriculture pilot project using *damsissa*, *Ambrosia maritima*, as a molluscicide began in Munufiya governorate in February 1997. We were told that the plan was to plant *damsissa* by the side of canals. After six months, the plants were to be trimmed, and the cut branches thrown into the canal. This project also included the Snail Control Section of the Schistosomiasis

Control Program, which was to identify snail-infested areas suitable for planting *damsissa*, and was later to monitor the program to see if snails are present in treated areas, and if any are infected. Pilot projects that have been undertaken elsewhere in the Delta show that the application of dried *damsissa* resulted in marked declines in snail density (Allam 2000; Barakat et al. 1993). As this plant grows wild by the banks of canals in Upper Egypt, transplanting it to northern Egypt would be far cheaper than using petroleum based chemicals. Judging from experiments elsewhere it would also be less toxic for other canal life.

Toward safe water and sanitation

Safe water and sanitation play an important part in the maintenance of good health, and, as we have seen, can play their part in preventing schistosomiasis transmission. However, to be effective the systems must be properly maintained and used. What kind of changes could facilitate the maximum use of these facilities, or put another way, how could some of the barriers to effective use be removed?

As we have seen, the criteria by which scientists judge the safety of water for drinking and domestic use are not the same as the criteria applied by rural women in deciding which water source is appropriate for their various needs. As we saw in chapter 8, the water in the piped supply that reaches individual households in most rural areas of the Delta is drawn from subsurface sources. This water has a high content of salts that render it "hard." In our study communities, al-Garda and al-Salamuniya, we found that many women said that this hard water did not dissolve the soap, and left their clothes looking dull and their dishes less than shiny. This was one of their principal arguments for continuing to use the canals for their household tasks. If a way could be found to remove these salts, at a reasonable cost and near the source of the supply, fewer women would be likely to use the canals for domestic activities. An alternative would be to draw water from surface sources such as canals, but this would require expensive treatment to meet safe water standards.

This brings us to the second element in any consideration of how to ensure the use of safe water: cost. As we saw in chapter 9, in al-Garda and al-Salamuniya the high cost of metered water supplies, and the additional 50% charged (on top of the water bill) for a sewerage connection had discouraged women from making full use of local safe water supplies. In 1991, the subsidized charge was between 10 and 15 piasters per cubic meter, but actual production costs were approximately 50 piasters per cubic meter (K-Konsult 1991: 64). By 1993–94 the government was beginning to remove subsidies, to satisfy the structural adjustment program imposed by the International

Monetary Fund and the World Bank. If subsidies were completely removed, the basic cost of water would thus increase five-fold, with a corresponding five-fold increase for a sewerage connection.

The IMF and its supporters made two arguments in favor of removing subsidies: firstly the ideological requirement that users should pay the full price for services, and secondly, the need to increase income in preparation for the eventual privatization of the public water and sewerage utilities. In 1991, consultants presenting a water and sanitation plan for Munufiya governorate considered that the increased income derived from the metered water supply (in contrast to the previous flat rate per household) would help to raise funds for maintenance and for new projects. It would also make people aware of the value of water and prevent its misuse. They reported that metering helped to keep the per capita consumption in rural areas without sewerage as low as 30 to 70 liters a day. Average water consumption may be as low as 41 liters per capita where water supply is limited to one or two taps and there is no sewer system (K-Consult 1991: 64).

The consultants made no mention of the usually quoted minimum per capita consumption level necessary to maintain good health. This minimum is 40 liters per day for all purposes—drinking, cooking, hygiene, and sanitation—when people use *public taps*. Further health benefits result when people have household connections, and in such cases the minimum requirement is expected to increase (WSSCC 2000: 35). The consultants quoted with apparent approval a consumption level of 30 or 40 liters per day by people with *household* connections.

Egyptians have long regarded water as a "free good," as a gift from God to be shared by all; metering a water supply deters people from sharing water with their neighbors. Yet women do not waste water, as the consultants imply. Some women still reuse water, even if they no longer have to bring it into the house from a public tap. As we saw in al-Garda, some local housewives who were connected to the water and sewerage system were still reusing water. As a result, the domestic wastewater that entered the Aqualife sewage system had chemical and bacteriological characteristics more like sewage than sullage; this was likely to be the main reason why the Aqualife system was not working properly.

Consultants and government officials are acutely aware of an impending national water shortage, as virtually all Egypt's water is drawn from the River Nile and its aquifer, supplemented by "fossil" underground water supplies originating in earlier periods of high rainfall. However, using pricing as a mechanism to conserve water could, at some point, result in poor people using insufficient water for health and hygiene requirements, as well as forcing them to use alternative unsafe sources, such as canal water. However they are financed, safe water supplies must be affordable for users. In order to move forward toward the creation of a safe water and sanitation system which rural people are eager and willing to use at all times, a balance must be achieved between the input of government planners and foreign consultants and the real world needs of rural Egyptians.

A major reason why women in al-Garda and al-Salamuniya washed at the canal rather than in their own homes was the difficulty of removing domestic wastewater from the house after washing. This problem could be solved by an effective, safe sanitation system. Such a system would also prevent the contamination of subsurface water from sewage holding tanks and the release of untreated sewage (possibly containing schistosome eggs) into surface water such as canals.

In the 1980s and 1990s, Egyptian engineers were working to identify affordable and sustainable sanitation systems for large rural settlements, such as small bore systems linked to existing holding tanks, that would remove both wastewater and sewage (Watts and El Katsha 1995; Gemmell et al. 1991). These would be less costly to install and maintain than the more technically complex system in al-Garda.

In 1995 we met with residents in al-Salamuniya and an engineer from the High Institute of Public Health in Alexandria to discuss some of the options available and likely costs. While villagers were enthusiastic about such a system, it was clear that funding would be required from outside the community.

In the absence of a sanitation system and in situations such as in al-Garda, where not all households were linked to the sewerage system, facilities to improve the safe collection of effluent and to control illegal dumping were urgently needed. In our discussions with residents, we found that they were acutely aware of this problem and were emphatic that existing fines should be enforced for illegal dumping of sewage and for piping wastewater and sewage directly into the canal. But, going beyond this, there was an urgent need for government authority to establish safe dumping places or treatment plants within a reasonable distance of each community.

On the household level, many disposal systems require upgrading and safe emptying procedures (although, in the long run, it is far safer to rely on a system in which sewage can be treated on site or removed by pipe to a treatment plant). A law introduced after the end of our field work states that all new houses must have a sealed tank for receiving domestic wastewater and sewerage, in order to avoid the contamination of subsurface water. However, this does not address the problem of emptying such tanks, whether they are in new or existing houses.

Solid waste collection

As of the early 1990s, al-Garda and al-Salamuniya did not have any facilities for collecting solid waste even though residents were well aware that their environment was being degraded by the practice of throwing garbage into the canals and the streets. In the baseline census, 70% of householders admitted that they threw rubbish into the canal. They told us that there was more solid waste now than in the past, especially now that non-biodegradable plastic bags and other containers were commonly used. Moreover, as most villagers now cooked on butagaz or on primus stoves, rather than in the old style mud brick oven, *furn*, they could no longer burn their household garbage. Farmers complained that, because there was now more junk of all sorts in the canals, they were forced to spend more time in the canals removing it, so that it would not clog up their irrigation pumps.

The Village Council has the legal authority to run efficient waste collection systems, and are able to collect fees up to 2% of the rental value of each house for this purpose. In 1987, in another Munufiya village, some of our team members worked with local people and their representative on the Village Popular Council to introduce a garbage collection system (El Katsha and Watts 1993: ch. 3). Nothing, however, has yet been done in al-Garda and al-Salamuniya.

Looking ahead: health service provisions

We are now taking stock of our findings, and of the changes have taken place in the Schistosomiasis Control Program since the completion of our research project in 1996. Firstly, we should note that the Ministry of Health has absorbed the Ministry of Population (responsible for family planning programs) and is now known as the Ministry of Health and Population (MOHP). We will begin by looking at the situation in the *curative* health services and then explore what has happened in the fields of community-based activities and health promotion.

Mass chemotherapy

When we were doing our research, the National Schistosomiasis Control Program was using a *selective* strategy for treatment, meaning they were providing praziquantel only for those who had been tested at a health unit and found to be positive. However, in 1997, based on the findings of many studies by the Schistosomiasis Research Project, the Ministry conducted a pilot program of mass chemotherapy, which subsequently became the mainstay of its program. This new approach treated everyone in highly endemic communities without prior testing.

By this time, praziquantel had been used long enough in Egypt to demonstrate that it was safe, well tolerated, and generally efficacious. Mass chemotherapy was also seen as cost effective, given that the drug was now produced locally and cost far less than it did in many other endemic countries. As of 1998 the cost was around 50 cents US for four tablets (the standard adult dose) compared to $ 4–5 US when it was first introduced (El Khoby et al. 1998).

In most instances, praziquantel continued to be effective for both forms of schistosomiasis. For example, a detailed study in a village in Giza governorate, just south of Cairo, found that after mass chemotherapy the prevalence rate of *S. haematobium* declined from 23.1% to 3.8% a year after mass treatment (Talaat and Miller 1998). In March-May 1997, mass chemotherapy against *S. mansoni* (the form of the disease that affects far more people in Egypt than *S. haematobium*), covered 10 million school children and residents of over 500 villages with high infection rates. After 6 months a reduction in prevalence of around 50% was recorded (El Khoby et al. 1998; Kusel and Hagen 1999). Official MOHP returns indicated that in 1999, in all Delta governorates, the level of *S. mansoni* overall was less than 5%, with fewer than 8% of school age children infected and fewer than 10% of outpatients (El Khoby et al. 2001: 19–20).

The achievements of the mass chemotherapy program have been impressive. But there are no published Ministry records indicating what proportion of the population in the highly endemic target areas, or among school children, was actually reached during these programs. Neither is there information about the gender of those treated. Such information could provide an indication of the completeness of coverage. The treatment of school age children was conducted through schools, thus it is likely that mass chemotherapy programs missed the out-of-school children, especially girls, who formed the most seriously affected group (see chapter 11). Also, we have no way of assessing how many women were not treated. This could have happened because they were pregnant, or simply because they were reluctant to come forward for mass treatment; as we have seen, women were less likely than men to seek treatment. Thus, the impressive figures we have for the coverage of mass chemotherapy may conceal a failure to reach certain groups, especially girls who were not in school and young married women.

Formal evaluations of the Ministry mass chemotherapy program indicated that some people who were treated were found to be positive after treatment. For example, in 1998, a second round of mass chemotherapy in the Delta resulted in a reduction in *S. mansoni* prevalence from around 30% to 7%, a reduction of approximately 75%. Gratifying though this figure was, it meant that 7% of the people were still excreting schistosome eggs. However, even if people became infected after treatment, or were not completely cured, they generally excreted fewer eggs than they had done before.

There are two reasons why people are found to be still infected after treatment: treatment failure (more common in *S. mansoni* than in *S. haematobium*) and reinfection. Overall, praziquantel has been estimated to fail to clear between 20 and 40% of infections, depending on age and intensity of infection. Also, as the drug is only effective against mature worms and eggs, it will not cure early infections; in such cases, young worms will survive to mature and expel eggs. (However, such infections may well be cured the next time round.)

Mass chemotherapy has successfully cured many of those infected, but, so long as people persist in going to the canals, reinfection will continue. In our study, reinfection appeared to be particularly likely among full-time and part-time farmers and among school-age children. With regard to school children, these findings are reinforced by a recent study of the impact of treatment for *S. mansoni* among school children in Kafr al-Sheikh governorate in the northern Delta, the governorate with the highest overall infection rates for schistosomiasis in Egypt. Successive annual mass treatment of children in the first five grades of primary school (between 6 and 11 years old) for five years, between 1996 and 2000, resulted in an overall reduction in infection rates from 70.9% to 38.6% for boys and from 66.7% to 36.2% for girls. Clearly, many of these children were being reinfected after treatment. Overall, the Geometric Mean Egg Count declined from 138 to 63 for boys and from 139 to 63 for girls (personal communication, Rashida Barakat and Hala El Morshedy, 2001).

These findings clearly show that mass chemotherapy alone cannot completely eradicate schistosomiasis. So long as reinfection occurs, the infection will be perpetuated in the canals flowing through rural areas. At the same time, certain pockets of infection (for example in small, remote settlements) will be missed, and certain sectors of the population are likely to be less thoroughly covered by treatment programs (such as pregnant women and children not in school), resulting in continuing sources of infection.

Praziquantel resistence and a future vaccine

Praziquantel has been widely used in Egypt since the late 1980s. During the years 1997, 1998, and 1999, over 67 million tablets and two million bottles of liquid formula (suitable for young children) were used (El Khoby and Fenwick 2001; Doenhoff et al. 2000). As a result, Egyptian physicians and researchers have been concerned that schistosomal strains resistant to the drug may develop and threaten the effectiveness of the control program. This is important in view of the growing evidence of the resistance of widespread diseases such as malaria and TB to formerly effective antibiotics and other drugs (Hall 2000; Farmer 2001: ch.8). This resistance has been associated with changes in the disease agent caused in part by over-prescribing or from patients' failure to complete the recommended full course of treatment. At the moment, the evidence for resistance to praziquantel is less than fully convincing. One reason for this is that it is very difficult to identify the specific causes of treatment failure in field studies. Researchers also disagree about the significance of evidence from laboratory studies of mice infected with various schistosome strains (Utzinger et al. 2000; Ismail et al. 1999; Kusel and Hagen 1999). However, alternatives to praziquantel are available, especially metrifonate (for *S. haematobium*) and oxamniquine (for *S. mansoni).* WHO has recommended that these alternative drugs should be kept in production in case they are needed (Gryseels 2000; Kusel and Hagen 1999).

Concern about resistance to praziquantel adds urgency to the search for a vaccine that would prevent people becoming infected in the first place. So long as there is no change in human behavior associated with exposure to, and contamination of, canals, a vaccine appears to be a "magic bullet," a strategy of proven effectiveness against childhood infectious illness, and a central plank in the global eradication of smallpox (in 1976). However, recent research suggests that it will be many years before an effective vaccine against schistosomiasis can be developed and that there would be serious ethical problems involved in testing it on human subjects (Gryseels 2000; Bergquist and Colley 1998; Wilson and Coulson 1998). Thus, an integrated strategy based on praziquantel and incorporating a range of preventive strategies remains the only feasible approach to schistosomiasis control for the foreseeable future.

The role of the rural health units

Mass chemotherapy has been shown to result in a dramatic fall in infection rates, even though it may miss certain sectors of the population. But, inevitably, a point will be reached when mass therapy will no longer be cost effective. At this stage, selective screening and treatment, based at government health units, will resume its former major role in schistosomiasis control. Thus, the local health facilities, especially rural health units, will once again play a dominant role in the schistosomiasis control program. In the face of the increasing demand for income generation in the public health sector, it is especially important that screening and treatment should continue to be available free to all. Egyptians currently make good use of these services, but we do not know how many of them would continue to use them if they had to pay.

People in rural areas need to know that they should still go to health centers for free diagnosis and treatment. To meet this challenge, staff training programs for curative as well as preventive strategies need to be developed to include gender sensitivity and communication skills, especially for face-to-face outreach in the community setting. Our research has demonstrated that unless testing procedures are gender sensitive they will miss groups such as girls (both in and out of school) and young women. The revised strategy for school-based screening that we developed and tested (see chapter 11) could form the basis for such an effort. Diagnostic and treatment strategies for public health facilities could also be reviewed with a view to targeting vulnerable groups and inaccessible areas.

Surveillance of the local population to identify those who need retesting after treatment, or require targeted attention for treatment and health promotion, is an essential part of control activities. Records of those treated could be entered in the recently introduced family health record cards so that the patients can be followed up three months later. Health staff need access to records of the number of people tested and treated, according to gender, in their facility. Any marked gender inequality, or change in the ratio of males and females tested or treated could alert staff to their possible failure to reach a certain group or groups and their need to increase their efforts to reach them. Similarly, in order to improve services at all levels, all data produced by the district and governorate, and the central MOHP office in Cairo, should be disaggregated according to gender.

The RHU can provide a focal point for face-to-face community based activities, supported by the primary health care services, and involving schools. Such activities are complementary to the use of mass media, especially TV, which regularly carries messages about schistosomiasis. They can target information for behavior change to specific groups, recognizing their different sources of exposure to infection and variations in access to screening and treatment. The design of such programs needs to recognize the gendered behavior of different target groups (each of which is affected in varying degrees by knowledge and distinctive attitudes).

What we are talking about here involves capacity building. This requires on-going training and follow-up, face-to-face targeted interactions between health providers and other local staff and local residents. During the mid-1990s when we were conducting our research, we found that these approaches played only a small role in formal schistosomiasis control activities.

The role of the MOHP

The Ministry of Health and Population is aware that after mass chemotherapy ceases to be a viable policy, schistosomiasis could again emerge as a major disease threat in certain rural areas. At this stage there will likely be small pockets of infection in remote areas, as well as certain population groups which have been neglected and which could contribute to the future expansion of the disease. We have seen examples of what has happened in some parts of the world with diseases that health authorities thought had been successfully controlled. For various reasons, including drug resistance and the failure to sustain surveillance programs, diseases such as malaria, sleeping sickness, and TB have once again become serious public health problems after being considered effectively controlled in most endemic areas in the 1960s and 1970s (White et al. 1999; Burri 2001; Farmer 2000: ch. 8). Clearly, each of these diseases has a specific transmission pattern associated with specific human activities, and a range of social and political settings, but in all these cases policy makers became complacent and were caught off guard by a rapidly changing situation.

There is also the issue of "staying power," or rather its absence. Carried away by new ideas, and the need to make their own distinctive professional mark, policy makers and managers are all too often tempted to move on to exciting new projects, and to become less thorough in performing standard, routinized—but extremely necessary—tasks such as surveillance.

It has recently been pointed out that schistosomiasis often persists as a problem years after transmission has ceased and that a new phenomenon known as "post-transmission schistosomiasis" needs to be recognized. From our point of view, this phenomenon is most important because individuals, who can harbor the eggs of *S. mansoni* for up to twenty-five years, will continue to excrete schistosome eggs in feces and, possibly, initiate a new transmission cycle in local canals. At the same time, tissue damage caused by schistosome eggs can continue to develop into cancers or other life-threatening conditions that the public health services have to diagnose and treat at great expense (Editorial 2000). For these reasons, even though its public profile will be lower, the MOHP will necessarily continue to be concerned about schistosomiasis control, knowing that, if it lets up its guard, schistosomiasis may well flare up again.

Enabling schistosomiasis control: the Egyptian health care system

In looking at the *Egypt Human Development Report 1997/8*, we find the following statement: "The Egyptian health care system is characterized by a low level of coordination between its various organizations. This results in fragmentation of the system, which has a negative impact on its performance." (EHDR 1998: 62) Our research findings illustrate vividly the validity of this finding, as it applies to our local level study of the delivery of services at the local level.

The *Egypt Human Development Report 1997/8* also comments that: "The main inefficiency areas in the Egyptian health sector originate mainly, if not exclusively, from the under-utilization of its potentials." (EHDR 1998: 62) The interventions that we designed and tested included a system of testing stool samples for school children, and a revision of the recording sheets. We emphasized from the beginning that these could be carried out using the existing staff and resources available at a fairly typical Rural Health Center.

In addition to unused potential, other barriers to efficiency in the health system mentioned in the *Egypt Human Development Report*, such as the physical deterioration of facilities and shortage of supplies are, as of the year 2000, being addressed by the Ministry. One innovation we noted since we completed our research was the designation of specific staff in RHUs as "health educators." These and other changes should contribute to an improvement in schistosomiasis related services.

The main strategies for schistosomiasis control delivered through the rural health units are the same in the Nile delta and in the river valley to the south. While the main issues involved in each intervention are the same, each region and sub-region is different and may require a different mix of control strategies (Mehanna et al. 1994). Our research and interventions were carried out in a relatively wealthy area of rural Lower Egypt, where levels of literacy and school attendance were high. In other areas, strategies may need to be modified, as a result of inquiries into local conditions and listening to local people, to make sure that the activities are appropriate for the local setting. For example, in areas with a scattered population (such as in newly settled areas on the fringe of the Delta), the diagnostic and treatment program provided by the health facility may need to be supplemented by mobile clinics. In areas where many children are not in school, alternative strategies will be needed to reach this group. Educational messages may have to be targeted to specific groups not covered in an initial plan, or modified for areas where specific risk behaviors have been identified. Giving local health authorities a greater flexibility and fostering health promotion and local involvement would help to identify and meet these local needs, within a framework that identified national priorities. The current stress on decentralizing the Ministry of Health, and other relevant ministries, provides such an opportunity.

Looking to the future, Sohair Mehanna and Peter Winch (1998) have suggested two alternative paths for rural health units in Egypt. They could maintain the current system in which the public health service provides for a wide range of services for everyone, or they could relinquish most of these tasks to the private sector. If the second option was taken, we suggest that schistosomiasis control must necessarily still remain within the sphere of the public health service, and that this service must offer free diagnosis and treatment. Rural health units should also continue to play a large role in preventive services such as health promotion. Therefore, we see a continuing role for local public health facilities in the treatment of schistosomiasis, and especially in more broadly based prevention activities.

Enabling schistosomiasis control: Communication processes

We have earlier suggested that local residents' groups and networks cannot provide for all local needs; government agencies are needed to finance local improvements, and to maintain them. The 1960 local government reforms established Village Councils, as the lowest tier of government, to deliver and maintain local services, such as water and sanitation. The Popular Village Council, consisting of elected members, was established to act as a conduit for the requests, or complaints of local people about such services and their maintenance to the Executive Village Council, which consists of permanent government employees responsible for the various sectoral activities.

In practice, however, there is a communication gap between the staff who work for the Village Council and the local residents. These local staff lack the communication skills and information to respond effectively to local needs (as indeed do the staff at the health units). Without a responsive local staff, and guidelines to encourage them to meet local needs, residents getting together to improve conditions in their community, or requesting health information, are all too often facing a brick wall. This is a common failing in a centralized system, where information travels upward and directives travel downward, with little effective feedback. This means that staff spend a lot of time responding to the demands of their superiors (filling in forms etc.) and are not encouraged to meet the needs of local people who are supposed to be the beneficiaries of their services.

This lack of communication between local government and rural citizens was identified in an earlier study in Munufiya governorate carried out by some of the members of our schistosomiasis research team. During this action research project with local women, it emerged that a major problem in implementing repairs and maintenance for standpipes and installing safe handpumps was that district and local staff working for the Village Council did not know how to meet the needs of local people (El Katsha and Watts 1993).

Thus, a follow-up research project, lasting from April 1992 to October 1994, was designed to explore ways of developing an effective partnership between decision-makers, local level government staff, and community residents. This study focused on developing networking strategies and communications processes at all levels and by all actors, at the village, the *markaz* (district) and the governorate level, and identifying the role all partners played in sustaining health and sanitation projects.

One of the major outcomes of this project was a manual, in Arabic, on village development, that identified the conceptual framework of the communication processes—the steps and methodologies that were undertaken by the partners to identify problems that could be solved, and what their various tasks should be. The researchers conducted workshops in which participants developed a plan of action for the two villages, identifying what was needed, how to mobilize support for these activities, and how to plan and implement them (El Katsha and Badran 1994).

Training in such communication skills would certainly help to initiate environmental improvements in communities such as al-Garda and al-Salamuniya. It would also be helpful if enabling regulations were provided for local input into planning processes, for example for communal waste disposal systems and communal laundries.

Prevention and the need for intersectoral collaboration

A holistic approach to schistosomiasis control requires collaboration between the Ministry of Health and Population and the ministries responsible for education, water supply, environment, sanitation, housing, irrigation, agriculture, and local government. Yet collaboration seems to be difficult in a government structure that is still highly centralized and vertically organized.

We found some examples of what could happen when there is no communication between the MOHP and other ministries. For example, during the 1990s, as part of the development of an improved system of water management by farmers at the *misqa* level, the Ministry of Public Works and Water Resources began to implement a policy of continuous flow of water through the canal system, including the *misqa*s. Following the relaxation of central control of farmers' cropping patterns in the mid-1980s, the farmers needed regular access to water to enable them to meet the water demands of the crops they had decided to grow (see Hvidt 1988, chapter 1). In view of the central role played by canal systems in the transmission of schistosomiasis, any change in canal regimes should require collaboration between the ministries responsible for irrigation and health. In 1993, WHO identified the urgent need for health authorities to negotiate with agencies responsible for water, irrigation, and agriculture, as well as various non-government agencies, in order to develop an overall health plan for water resource projects and to include health in environmental assessments (Hunter et al. 1993, chs. 8 and 9).

Since the early-1990s environmental issues have been given a greater prominence in Egyptian government activities and in public discussions. The Ministry of Environment, founded in 1997, absorbed the Egyptian Environmental Affairs Agency. It was given funds and legal authority to administer Law 4 for 1994, which regulates and enforces measures against water and air pollution (Hopkins and Mehanna 2000: 4–5). The relationship between schistosomiasis infection and environmental conditions relating to canals, safe water, and sanitation prompts us to suggest the value of closer contact between the MOHP and the Ministry of State for Environmental Affairs. Without such collaboration there will be more embarrassing incidents such as one that occurred shortly after the end of our project. On this occasion, the MOHP, in association with a mass chemotherapy program, sent out a directive that canals in these areas should be chemically treated. In a community near one of our study villages, the snail inspectors followed instructions, using remaining supplies of copper sulfate (rather than niclosamide) and were promptly arrested for polluting the canals by officials from the Egyptian Environmental Affairs Agency.

Intersectoral collaboration is essential if prevention is to play a greater part in integrated control programs than it did in the mid-1990s, at the time of our research. Prevention strategies are complex and many-faceted, partly because they draw on expertise in so many areas—public and environmental health, health promotion, civil and irrigation engineering, housing, and education as well as the insights of politics, economics, cybernetics, management, and planning. They also draw on social scientists' insights into the social and cultural setting in which human behavior takes place.

Preventive strategies are cost effective, in so far as they bring about sustainable behavioral change and long term improvements in the environment (in terms of added health benefits, less arduous daily work for farmers, women, and girls, and a pleasanter environment). However, prevention is not a "quick fix"; it does not make headline-grabbing news on the nightly TV reports or in newspapers. It takes time to improve the environment, and to bring about the many behavioral changes that have an effect on health. The changes are incremental and inherently undramatic. But they are real, and sustainable, and benefit all sections of the community.

Appendix: Methodology

A. Criteria for village selection
B. Epidemiology
C. Short Surveys
D. Group and individual interviews
E. Focus group discussions—topics, participants, and guide lists.

A. Criteria for village selection

One village with piped water only, a second also with a sewerage system.

The villages should be:

- of approximately equal size
- similar in social and occupational structure
- accessible from Cairo
- accessible to primary health care services in or near the village, and with a health staff willing to participate in the project
- in a district that had district level staff willing to participate in the project
- villagers favorable to the activities of the research team
- endemic for schistosomiasis

B. Epidemiology

SRP testing protocol:

A single specimen was taken from each individual, from which two slides were prepared. Both techniques provided for a quantitative measure of egg output, intensity, rather than simply identifying eggs in the sample provided.

A 10% quality control of all samples collected during the project was conducted in the lab of the High Institute of Public Health in Alexandria.

Sampling:

To obtain valid findings from the epidemiological survey, around one hundred positive cases of *S. mansoni* would be needed in each community. A 5% sample of villagers was therefore tested for both types of schistosomiasis. The results are shown in table A.1. Based on the findings of the pilot survey, this would require a random sample of 15% from al-Garda (approximately 116 expected cases from 1156 individuals), and a 6% sample in al-Salamuniya (around 129 cases from 517 individuals).

Table A.1: Results of parasitological tests of 5% sample in al-Garda and al-Salamuniya, 1991

	Sample	S. mansoni	S. haematobium
al-Garda	362	33, 9.1%	1 (0.3%)
al-Salamuniya	406	103, 25.4%	1 (0.3%)

Following these guidelines, the first-round survey in al-Garda provided 992 samples of which 80 were positive (8%); al-Salamuniya provided 454 samples, of which 112 were positive (24%). The final database excluded the small *'izba*, hamlet, close to al-Salamuniya, that had been originally included in the sample.

C. Short surveys

A number of short surveys were designed incorporating open-ended questions to provide a greater depth to the responses.

i. 80 households, all those containing individuals who tested positive in 1992, with a total of 226 adults, were asked to identify, and name, any places were they had contact with canal water, and what activities they carried out there. Other questions included knowledge of schistosomiasis, its severity and treatment, and why people persisted in using canals. It ended with open-ended questions about what could be done to improve the village environment.

ii. 12 school teachers—about knowledge and training in schistosomiasis (preliminary to Focus Group Discussions).

D. Interviews and group discussions

These involved:

i. MOH staff—to identify schistosomiasis control procedures and their perceptions of barriers to implementation.

ii. School teachers—about opportunities for outreach to schoolchildren.

iii. Irrigation staff—possible changes in the irrigation regime that might affect the habitat of vector snails.

iv. Local Village Council staff—for insights into water and drainage/sewerage conditions.

v. Villagers—about defecation practices.

E. Focus group discussions

i. Topics and composition.

Table A.2: Topics and composition of focus group discussions

Topics	Groups	# participants	Gender	# meetings
Agriculture and exposure to schistosomiasis	Farmers, adult men, and youths	15	M	2
Knowledge of schistosomiasis	Men and youths	17	M	2
Knowledge of schistosomiasis and canal use	Young and elderly women	12	F	2
Knowledge of schistosomiasis and canal use	Females < 20 yrs	15	F	2
Knowledge of schistosomiasis and role at RHU	All RHU staff in al-Salamuniya	20	M & F	1
Knowledge of schistosomiasis, its source, and canal use	3rd and 4th grade schoolchildren	11	M	2
		13	F	2
Knowledge of schistosomiasis	Teachers	15	M & F	1

ii. Focus Group Discussions Guidelines
Guide questions for focus group discussions held in al-Garda and al-Salamuniya in April 1992.

Women's Groups:

1. For what purpose do you use the canals?
   - why?
   - at what time of the day?

2. Is schistosomiasis dangerous?
   - is it contagious?
   - is everyone liable to get it?
   - which groups are most liable to get it?
   - how can they get it?
   - what are the symptoms?
   - how can you find out if you have it?
   - what is the treatment?

3. If you know all the above, why do you continue using the canals?

4. What pollutes your canal?

5. Can we, as villagers, help to prevent pollution?

6. Where do you get your information about schistosomiasis?

7. Did this information make you change your behavior?

Men:
As for women, but greater concentration on the following agricultural questions:

1. What are the agricultural chores which need water contact?
    - during which seasons does this contact occur?
    - which crops are most involved?

2. What age group is involved most in irrigation activities?

3. Detailed description of exactly what they do during irrigation which results in water contact.

4. Do new irrigation systems result in less or more water contact?

5. Where do you defecate and urinate in the fields?

Bibliography

Abdel-Wahab M. F. 1982. *Schistosomiasis in Egypt.* CRC Press: Boca Raton, Florida.

Abdel-Wahab, M. Farid et al. 2000A. The epidemiology of schistosomiasis in Egypt: Fayoum Governorate. *American Journal of Tropical Medicine and Hygiene.* 62 (2), Supplement: 55–64.

Abdel-Wahab, M. Farid, Gamal Esmat, Eman Medhat, Shaker Narooz, Iman Ramzy, Yasser El-Boraey and G. Thomas Strickland. 2000B. The epidemiology of schistosomiasis in Egypt: Menofia Governorate. *American Journal of Tropical Medicine and Hygiene.* 62 (2), Supplement: 28–34.

Abdel-Wahab, M. Farid, Gamal Esmat, Yasser El-Boraey, Iman Ramzy, Eman Medhat and G. Thomas Strickland. 2000C. The epidemiology of schistosomiasis in Egypt: methods, training, and quality control of clinical and ultrasound examinations. *American Journal of Tropical Medicine and Hygiene.* 62 (2), Supplement: 17–20.

Abdel-Wahab, M. F., Ayman Yosery, Shaker Narooz, Gamal Esmat, Salih El Hak, Samir Nasif and G. T. Strickland. 1993. Is *Schistosoma mansoni* replacing *S. haematobium* in the Fayoum? *American Journal of Tropical Medicine and Hygiene.* 49: 697–700.

Abdel-Wahab, M. F., G. T. Strickland, A. El-Sahly, N. El-Kady, Sohair Zakaria and Laila Ahmed. 1979. Changing pattern of schistosomiasis in Egypt 1935–1979. *Lancet.* 2 (8136): 242–44.

Akogun, Oladele B. 1991. Urinary schistosomiasis and the coming of age in Nigeria. *Parasitology Today.* 7: 62.

El Alamy, M. A. and Barnett L. Cline. 1977. Prevalence and intensity of *Schistosoma haematobium* and *S. mansoni* infection in Qalyub, Egypt. *American Journal of Tropical Medicine and Hygiene.* 26: 470–72.

Ali, Kamran Asdar. 1998. Conflict or cooperation: changing gender roles in rural Egyptian households. In Hopkins & Westergaard, *Directions of Change in Rural Egypt.* Ch. 8 (166–83)

Allam, Amal Farahat. 2000. Evaluation of different means of control of snail intermediate host of *Schistosoma mansoni. Journal of the Egyptian Society of Parasitology.* 30: 441–50.

Allen, Louise F. 1989. *The Situation of Children in Upper Egypt.* Central Agency for Public Mobilization and Statistics (CAPMAS) and United Nations Children's Emergency Fund (UNCEF): Cairo.

Assaad, M. and S. El Katsha, 1981. Villagers' use of and participation in formal and informal health services in an Egyptian delta village. *Contact.* 65: 1–7.

Barakat, Rashida, Azza Farghaly, A. G. El Masry, Medhad K. El Sayed and Mohamed H. Hussein. 2000. The epidemiology of schistosomiasis in Egypt: patterns of *Schistosoma mansoni* infection and morbidity in Kafr El-Sheikh. *American Journal of Tropical Medicine and Hygiene.* 62 (2), Supplement: 21–27.

Barakat, R., A. Farghaly, M. F. El-Sawy, N. K. Soliman, J. Duncan, A. Zaki and F. D. Miller. 1993. An epidemiological assessment of *Ambrosia maritima* on the transission of schistosomiasis in the Egyptian Nile delta. *Tropical Medicine and Parasitology.* 44: 181–86.

Barker, Carol and Andrew Green. 1996. Opening the debate on DALYs. *Health Policy and Planning.* 11: 179–83.

Bergquist, R. N. and D. G. Colley. 1998. Schistosomiasis vaccines: research to development. *Parasitology Today.* 14: 99–104.

Berman, Peter, Carl Kendall and Karabi Bhattacharyya. 1994. The household production of health: integrating social science perspectives on micro-level health determinants. *Social Science and Medicine.* 38: 205–15.

Blumenthal, U. J. 1989. Assessment of Human Water Contact Patterns—Issues of Data Collection. Paper Presented at Workshop on Women in Tropical Diseases, 11–17 October 1989, Cairo, Egypt.

Boot, Marieke T. and Sandy Cairncross (eds.). 1993. *Actions Speak: the Study of Hygiene Behaviour in Water and Sanitation Projects.* IRC International Water and Sanitation Center and London School of Hygiene and Tropical Medicine.

Bourne, Peter (ed.). 1983. *Water and Sanitation: Economic and Sociological Perspectives.* Academic Press: Orlando, Florida.

Brieger, William. 1994. Measuring hygiene behavior in rural Nigeria. In Cairncross, Sandy and Vijay Kuchar (eds.), *Studying Hygiene Behavior: Methods, Issues and Experiences.* Sage Publications: New Delhi and London. 202–9

Brown, Peter J, Marcia C. Inhorn and Daniel J. Smith. 1996. Disease, ecology, and human behavior. In Sargent, Carolyn F. & Thomas M. Johnson (eds.), *Medical Anthropology: Contemporary Theory and Method.* Praeger: Westport, Connecticut. 183–218.

Bundy, D.A.P. and U. J. Blumenthal. 1990. Human behaviour and the epidemiology of helminth infections: the role of behaviour in exposure to infection. In C. J. Barnard and J. M. Behnke (eds.), *Parasitism and Host Behaviour.* Taylor and Francis: London. 264–89.

Bundy, Donald A. P. and Helen L. Guyatt. 1996. Schools for health: focus on health, education and the school-age child. *Parasitology Today.* 12: 1–16.

Burri, Christian. 2001. Are there new approaches to roll back trypanosomiasis? *Tropical Medicine and International Health.* 6: 327–29.

Cairncross, Sandy, Ursula Blumenthal, Peter Kolsky, Luiz Moraes and Ahmed Tayeh. 1996. The public and the domestic domains in the transmission of disease. *Tropical Medicine and International Health.* 1: 27–34.

Chambers, Robert. 1997. *Whose Reality Counts? Putting the First Last.* Intermediate Technology Publications: London.

Cheesmond Ann K. and Alan Fenwick. 1981. Human excretion behaviour in a schistosomiasis endemic area of the Gezira, Sudan. *Journal of Tropical Medicine and Hygiene.* 84: 101–7.

Chitsulo, L., P. Engels, A. Montresor and L. Savioli. 2000. The global status of schistosomiasis and its control. *Acta Tropica.* 77: 41–51.

Clark, William D., Paul M. Cox, Lynne H. Ratner and Rafael Correa-Coronas. 1970. Acute *Schistosomiasis mansoni* in 10 Boys: an outbreak in Caguas, Puerto Rico. *Annals of Internal Medicine.* 73: 379–85.

Cline, Barnett L. 1995. The slow fix: communities, research and disease control. *American Journal of Tropical Medicine and Hygiene.* 52: 1–7.

Cline, B. L., M. Habib, F. Gamil, F. Abdel-Aziz and M. D. Little. 2000. Quality control for parasitologic data. *American Journal of Tropical Medicine and Hygiene.* 62 (2), Supplement: 14–16.

Cline, B. L. and B. S. Hewlett. 1996. Community-based approach to schistosomiasis control. *Acta Tropica.* 61: 107–19.

Cline, Barnett L, Frank O. Richards, M. A. El Alamy, S. El Hak, Ernesto Ruiz-Tiben, Janet M. Hughes and David F. McNeeley. 1989. 1983 Nile delta Schistosomiasis Survey: 48 years after Scott. *American Journal of Tropical Medicine and Hygiene.* 41: 56–62.

Collins, K. J. et al. 1976. Physiological performance and work capacity of Sudanese cane cutters with *Schistosoma mansoni* infection. *American Journal of Tropical Medicine and Hygiene.* 25: 410–21.

Coreil, Jeannine. 1995. Group interview methods in community health research. *Medical Anthropology.* 16: 193–210.

Doenhoff, M. J., G. Kimani and D. Cioli. 2000. Praziquantel and the control of schistosomiasis. *Parasitology Today.* 16: 364–66.

Dunn, Frederick L. 1979. Behavioural aspects of the control of parasitic diseases. *Bulletin of the World Health Organization.* 57: 499–512.

Editorial. Post-transmission schistosomiasis: a new agenda. *Acta Tropica.* 77: 3–7.

EDHS. 1995. *Egypt Demographic and Health Survey 1995.* National Population Council, Cairo, Egypt and Macro International Inc., Calverton, Maryland.

EHDR. 1998. *Egypt Human Development Report 1997/8.* Institute of National Planning: Cairo.

Elgood, Bonté Sheldon. 1908. Bilharziasis among women and girls in Egypt. *The British Medical Journal.* II (October 31): 1355–57.

Ernould J. C., K. Ba and B. Sellin. 1999. The impact of the local water-development programme on the abundance of the intermediate hosts of schistosomiasis in three villages of the Senegal River delta. *Annals of Tropical Medicine and Parasitology.* 93: 135–45.

Esry, Steven A. 1996. Water, waste and well-being: a multicountry study. *American Journal of Epidemiology.* 143: 608–23.

Esry, S. A., J. B. Potash, L. Roberts and C. Shiff. 1991. Effects of improved water supply and sanitation on ascariasis, diarrhoea, dracunculiasis, hookworm infection, schistosomiasis, and trachoma. *Bulletin of the World Health Association.* 69: 609–21.

Farley, John. 1991. *Bilharzia: A History of Imperial Tropical Medicine.* Cambridge University Press: UK.

Farmer, Paul.1999. *Infections and Inequalities: The Modern Plagues.* University of California Press: Berkeley and Los Angeles, California.

Farooq, Mohamed and M. B. Mallah. 1966. The behavioral pattern of social and religious water-contact activities in the Egypt-49 Bilharziasis Project Area. *Bulletin of the World Health Organization.* 35: 377–87.

Farooq, M., J. Nielsen, S. A. Saman, M. B. Mallah and A. A. Allam. 1996A. The epidemiology of *Schistosoma haematobium* and *S. mansoni* infections in the Egypt-49 Project area: 2. Prevalence of bilharziasis in relation to personal attributes and habits. *Bulletin of the World Health Organization.* 35: 293–318.

Farooq, M., J. Nielsen, S. A. Samaan, M. B. Mallah and A. A. Allam. 1996B. The epidemiology of *Schistosoma haematobium* and *S. mansoni* infections in the Egypt-49 Project area: 3. Prevalence of bilharziasis in relation to certain environmental factors. *Bulletin of the World Health Organization.* 35: 319–30.

Feachem, Richard G. 1984. Infections related to water and excreta: the health dimension of the decade. In Bourne, *Water and Sanitation*, 21–47.

Feldmeier, Hermann, Peter Leutscher, Gabriele Poggensee and Gundel Harms. 1999. Male genital schistosomiasis and haemospermia. *Tropical Medicine and International Health*. 4: 791–93.

Feldmeier, Hermann, Gabriele Poggensee, Ingela Krantz and G. Helling-Giese. 1995. Female genital schistosomiasis: new challenges from a gender perspective. *Tropical and Geographical Medicine*. 47 (2 Supplement): S2–15.

Feldmeier, Hermann, Gabriele Poggensee and Ingela Krantz. 1993. A synoptic inventory of needs for research on women and tropical parasitic diseases. II. Gender-related biases in the diagnosis and morbidity assessment of schistosomiasis in women. *Acta Tropica*. 55: 139–69.

Fenwick, A. and B. H. Figenschou. 1972. The effect of *Schistosoma mansoni* infection on the productivity of cane cutters on a sugar estate in Tanzania. *Bulletin of the World Health Organization*. 47: 567–72.

Frank, Christina et al. 2000. The role of parenteral antischistosomal therapy in the spread of hepatitis C virus in Egypt. *Lancet*. 355 (9207): 887–91.

Gabr, Nabil S., Tarek A. Hammad, Anwar Orieby, Eglal Shawky, Mahmoud A. Khattab and G. Thomas Strickland. 2000. The epidemiology of schistosomiasis in Minya governorate. *American Journal of Tropical Medicine and Hygiene*. 62 (2), Supplement: 65–72.

Gallagher, Nancy Elizabeth. 1993. *Egypt's Other Wars: Epidemics and the Politics of Public Health*. The American University in Cairo Press: Cairo.

Gemmell, J. S. 1989. Wastewater treatment in Egyptian rural development. *WasteWater International*. 4: 15–23.

Gemmell, J. S., O. El Sebaie, A. H. Gaber and P. G. Smith. 1991. Sullage disposal in rural Egypt – the degree of pollution and potential health hazards. *International Journal of Environmental Health Research*. 1: 183–91.

Gillies, Pamela. 1998. Effectiveness of alliances and partnerships for health. *Health Promotion International*. 13: 99–120.

Gomaa, Salwa Shaarawi. 1998. Political Participation of Egyptian Women. Final Technical Report to The American University in Cairo, project supported by UNICEF Cairo Office, and the Ford Foundation Regional Office in Cairo, May 1998.

Gryseels, B. 2000. Schistosomiasis vaccines: the devil's advocate's final plea. *Parasitology Today*. 16: 357–58.

Gryseels, B. 1994. Human resistance to *Schistosoma* infections: age or experience? *Parasitology Today*. 10: 380–84.

Gryseels, B. et al. 2001. Are poor responses to praziquantel for the treatment of *Schistosoma mansoni* infections in Senegal due to resistance? An overview of the evidence. *Tropical Medicine and International Health*. 6: 864–73.

Gundersen S. G. et al. 1996. Urine reagent strips for diagnosis of *Schistosomiasis haematobium* in women of fertile age. *Acta Tropica.* 62: 281–87.

El Hadidi, Haguer. 1990. *Sociocultural Factors Influencing the Prevalence of Diarrheal Disease in Rural Upper Egypt: an Ethnographic Study in Two Villages of Sohag.* UNICEF: Cairo and New York.

Hall, A. 2000. Mothers, malaria and resistance. *Tropical Medicine and International Health.* 5: 753–54.

El Hamamsy, Laila S. 1994. *Early Marriage and Reproduction in Two Egyptian Villages.* Occasional Paper, The Population Council/UNFPA: Cairo.

Hammad, Tarek A, Nabil S. Gabr, Maha M. Talaat, Anwar Orieby, Eglal Shawky and G. Thomas Strickland. 1997. Hematuria and proteinuria as predictors of *Schistosoma haematobium* infection. *American Journal of Tropical Medicine and Hygiene.* 57: 363–67.

Hammam, M., Farida A. Allam and Farouk Hassanein. 1975. Relationship between pure Schistosoma haematobium infection in Upper Egypt and irrigation systems. Part II: Host characteristics. The general prevalence of Schistosoma haematobium, age and sex distribution. *Gazette of the Egyptian Pediatric Association.* 23: 215–26.

Harb, M., R. Faris, A. M. Gad, O. N. Hafez, R. Ramzy, R. and A. A. Buck. 1993. The resurgence of lymphatic filariasis in the Nile delta. *Bulletin of the World Health Organization.* 71: 49–54.

Hopkins, Nicholas S. and Kirsten Westergaard (eds.). 1998. *Directions of Change in Rural Egypt.* American University in Cairo Press: Cairo.

Hopkins, Nicholas S., Sohair R. Mehanna and Salah El Haggar. 2001. *People and Pollution: Cultural Constructions and Social Action in Egypt.* American University in Cairo Press: Cairo.

Hopkins, Nicholas S. 1999. Irrigation in contemporary Egypt. In Bowman, Alan K. and Eugene Rogan (eds.), *Agriculture in Egypt from Pharaonic to Modern Times.* Proceedings of the British Academy 96, Oxford University Press. 367–85.

Huang, Yixin and Lenore Manderson. 1992. Schistosomiasis and the social patterning of infection. *Acta Tropica.* 51: 175–94.

Hunter, J. M., L. Rey, K. Y. Chu, E. O. Adekolu-John and K. E. Mott. 1993. *Parasitic Diseases in Water Resources Development: the Need for Intersectoral Negotiation.* World Health Organization: Geneva.

Husein, M. H., M. Talaat, M. K. El Sayed, A. El Badawi and D. B. Evans. 1996. Who misses out with school-based health programmes? A study of schistosomiasis control in Egypt. *Transactions of the Royal Society of Tropical Medicine and Hygiene.* 90: 362–65.

Hussein, Mohamed H., Medhat K. El Sayed, Maha Talaat, Amal El Badawy and F. DeWolfe Miller. 2000. Epidemiology 1, 2, 3: study and sample design. *American Journal of Tropical Medicine and Hygiene.* 62 (2), Supplement: 8–13.

Hvidt, Martin. 1998. *Water, Technology and Development: Upgrading Egypt's Irrigation System.* Tauris Academic Studies: London and New York.

Ibrahim, Barbara, Sunny Sallam, Sahar El Tawilla, Omaima El Gibaly and Fikrat El Sahn. 1999. *Transitions to Adulthood: A National Survey of Egyptian Adolescents.* The Population Council: New York and Cairo.

Inhorn, Marcia C. and Peter J. Brown (eds.). 1997. *The Anthropology of Infectious Disease: International Health Perspectives.* Gordon and Breach: Amsterdam.

Inhorn, Marcia C. 1995. Medical anthropology and epidemiology: divergences or convergences? *Social Science and Medicine.* 40: 285–90.

Ismail, Magdi et al. 1999. Resistance to praziquantel: direct evidence from *Schistosoma mansoni* isolated from Egyptian villagers. *American Journal of Tropical Medicine and Hygiene.* 60: 932–35.

Jordan, Peter. 2000. From Katayama to the Dakhla Oasis: the beginning of epidemiology and control of bilharzia. *Acta Tropica.* 77: 9–40.

Jordan, P., G. O. Unrau, R. K. Bartholomew, J. A. Cook and E. Grist. 1982. Value of individual household water supplies in the maintenance phase of a schistosomiasis control programme in Saint Lucia, after chemotherapy. *Bulletin of the World Health Organization.* 60: 583–88.

Joseph, Suad (ed.). 2000. *Gender and Citizenship in the Middle East.* Syracuse University Press: Syracuse, NY.

El Karim, M. A. Awad et al. 1980. Quantitative egg excretion and work capacity in a Gezira population infected with *Schistosoma mansoni. American Journal of Tropical Medicine and Hygiene.* 29: 54–61.

El Katsha, Samiha and Mohga Badran. 1994. Communication Processes: an Avenue for Sustaining Improved Health and Sanitation Practices. Final Project Report presented to the International Development Research Center, Ottawa.

El Katsha, Samiha and Susan Watts. 1998. Schistosomiasis screening and health education for children: action research in Nile delta villages. *Tropical Medicine and International Health.* 3: 654–60.

———. 1997. Schistosomiasis in two Nile delta villages: an anthropological perspective. *Tropical Medicine and International Health.* 2: 846–54.

———. 1995A. The public health implications of the increasing predominance of *Schistosoma mansoni* in Egypt: a pilot study in the Nile delta. *Journal of Tropical Medicine and Hygiene.* 98: 136–40.

———. 1995B. Schistosomiasis control through rural health units. *World Health Forum*. 16: 252–54.

———. 1994A. A model for health education. *World Health Forum*. 15: 29–33.

El Katsha, S. and S. J. Watts. 1994B. School-based summer clubs: venues for health education using a partnership model in Egypt. *Promotion & Education*. 1: 24–28.

El Katsha, Samiha and Susan Watts. 1993. *The Empowerment of Women: Water and Sanitation Initiatives in Rural Egypt*. Cairo Papers in Social Science. Volume 16, monograph 2. American University in Cairo Press: Cairo.

El Katsha, Samiha and Anne U. White. 1989. Women, water, and sanitation: household behavioral patterns in two Egyptian villages. *Water International*. 14: 103–11.

El Katsha, Samiha, Susan Watts, Amal Khairy and Olfat El Sebaie. 1993–94. Community participation for schistosomiasis control: a participatory research project in Egypt. *International Quarterly of Community Health Education*. 14: 245–55.

El Katsha, Samiha, Awatif Younis, Olfat El Sebaie and Ahmed Hussein. 1989. *Women, Water and Sanitation: Household Water Use in Two Egyptian Villages*. Cairo Papers in Social Science. Volume 12, monograph 2, American University in Cairo Press: Cairo.

Keatinge, H. P. 1927. *Records of the Egyptian Government Faculty of Medicine*. Government Press: Cairo.

Khairy, Amal E. M., Neguiba F. Loutfy, Ahmed Hamza and Mohamed D. El Borgy. 1986. Domestic water supplies and community self-help in Sidi-Ghazzi area, Nile delta—a strategy for schistosomiasis control. Part II, The incidence of schistosomiasis and the acceptability of available domestic water. *Bulletin of the High Institute of Public Health* (Alexandria). 16: 73–88.

Khalil, M. 1949. The National Campaign for the Treatment and Control of Bilharziasis from the Scientific and Economic Aspects. *Journal of the Egyptian Medical Association*. 32: 817–56.

Khalil, M. and M. Abdel Azim. 1935. The introduction of schistosoma infection through irrigation schemes in the Asswan area, Egypt. *Journal of the Egyptian Medical Association*. 18: 371–77.

Khattab, Hind Abou Seoud and Syada Greiss El Daeif. 1982. *Impact of Male Labor Migration on the Structure of the Family and the Roles of Women*. Regional Paper. The Population Council, West Asia and North Africa Region: Giza, Egypt.

Khattab, Hind, Nabil Younis and Huda Zurayk. 1999. *Women, Reproduction and Health in Rural Egypt: the Giza Study*. American University in Cairo Press: Cairo.

El Khoby, Taha. 2001. Morbidity control of schistosomiasis in Egypt. Paper prepared for presentation at the International Symposium on Schistosomiasis, Shanghai, China, July 2001.

El Khoby, T.A.G. 1995. Schistosomiasis control in a primary health care system. *Memorias do Instituto Oswaldo Cruz.* 90: 104.

El Khoby, Taha, Yehia Abdel Wahab, Mohamed Mostafa and Alan Fenwick. 2001. Egypt: A Case Study for the Control of Schistosomiasis. Report prepared for the Schistosomiasis Control Initiative, VACSERA, Agouza, Cairo, January 2001.

El Khoby, Taha, Mohamed H. Hussein, Nabil Galal and F. DeWolfe Miller. 2000A. Epidemiology 1, 2, 3: origins, organization, and implementation. *American Journal of Tropical Medicine and Hygiene.* 62 (2), Supplement: 2–7.

El Khoby, Taha et al. 2000B. The epidemiology of schistosomiasis in Egypt: summary findings in nine governorates. *American Journal of Tropical Medicine and Hygiene.* 62 (2), Supplement: 88–99.

El Khoby T., N. Galal and A. Fenwick. 1998. The USAID/Government of Egypt's Schistosomiasis Research Project (SRP). *Parasitology Today.* 14: 92–96.

El Khoby, Taha Abdel Gawad, Susan J. Watts and Samiha El Katsha. 1991. Schistosomiasis in Egypt: an update. *Nigerian Journal of Parasitology.* 12: 51–57.

King, C. L., F. D. Miller, M. Hussein, R. Barkat and A. S. Monto. 1982. Prevalence and intensity of *Schistosoma haematobium* infection in six villages of Upper Egypt. *American Journal of Tropical Medicine and Hygiene.* 31: 320–27.

Kloos, Helmut. 1995. Human behavior, health education and schistosomiasis control: a review. *Social Science and Medicine.* 40: 1497–1511.

Kloos, Helmut, Gene I. Higashi, Vernon D. Schinski, Noshy S. Mansour, K. Darwin Murrell and F. DeWolfe Miller. 1990. Water contact and *Schistosoma haematobium* infection: a case study from an Upper Egyptian village. *International Journal of Epidemiology.* 19: 749–58.

Kloos, Helmut, Gene I. Higashi, Jacqueline A. Cattani, Vernon D. Schlinski, Noshy S. Mansour and K. Darwin Murrell. 1983. Water contact behavior and schistosomiasis in an Upper Egyptian village. *Social Science and Medicine.* 17: 545–62.

Kloos, Helmut, Waguih Sidrak, Adly Abdel Malek Michael, Emad William Mohareb and Gene I. Higashi. 1982. Disease concepts and treatment practices relating to *Schistosomiasis haematobium* in Upper Egypt. *Journal of Tropical Medicine and Hygiene*. 85: 99–107.

K-Konsult. 1991. El Menoufeya Governorate Concise Feasibility Study for Water Supply and Sewerage. Interim Report prepared for NOPWASD, October 1991. Arab Republic of Egypt.

Kusel, J. and P. Hagan. 1999. Praziquantel – its use, cost and possible development of resistance. *Parasitology Today*. 15: 352–54.

Lane, Sandra. 1997. Television minidramas: social marketing and evaluation in Egypt. *Medical Anthropology Quarterly*. 11: 164–82.

Lane, Sandra D. and Afaf I. Meleis. 1991. Roles, work, health perceptions and health resources of women: a study in an Egyptian delta hamlet. *Social Science and Medicine*. 33: 1197–1208.

Lane, Sandra D. and Marcia Inhorn Millar. 1987. The "Hierarchy of Resort" reexamined: status and class differentials as determinants of therapy for eye disease in the Egyptian delta. *Urban Anthropology*. 16: 151–82.

Langsten, Ray and Kenneth Hill. 1994. Diarrhoeal disease, oral rehydration and childhood mortality in rural Egypt. *Journal of Tropical Pediatrics*. 40: 272–78.

Loza, Sarah F. 1993. Health education, mass media and schistosomiasis operations research results. Abstract of presentation at the Schistosomiasis Research Project International Conference on Schistosomiasis, Cairo, Egypt, February 14–18 1993, 41.

Madden, Frank Cole. 1907. *Biharziosis*. Cassell and Company: London.

Madden, Frank Cole. 1919. *The Surgery of Egypt*. Nile Mission Press: Cairo.

El Malatawy A., A. El Habashy, N. Lechine, H. Dixon, A. Davis and K. E. Mott. 1992. Selective population chemotherapy among schoolchildren in Beheira governorate: the UNICEF/Arab Republic of Egypt/WHO Schistosomiasis Control Project. *Bulletin of the World Health Organization*. 70: 47–56.

Manderson, Lenore. 1998. Applying medical anthropology in the control of infectious disease. *Tropical Medicine and International Health*. 3: 1020–27.

Mansour, N. S., G. I. Higashi, V. D. Schinski and K. D. Murrell. 1981. A longitudinal study of *Schistosoma haematobium* infection in Qena governorate, Upper Egypt. I. Initial epidemiological findings. *American Journal of Tropical Medicine and Hygiene*. 30: 795–805.

Mehanna, Sohair and Peter Winch. 1998. Health Units in rural Egypt: at the forefront of health improvement or anachronisms? In Hopkins and Westergaard, *Directions of Change in Rural Egypt*, 219–33

Mehanna, S., H. El Sayed, K. Dandash, S. Abaza and Peter Winch. 1998. Impact of school-based health education program on the control of schistosomiasis and other parasitic infections in rural Ismailia Governorate. *Zagazig University Medical Journal*. 4: 347–60.

Mehanna S., P. J. Winch, N. H. Rizkalla, H. F. El Sayed and S. M. Abaza. 1997. Factors affecting knowledge of the symptoms of schistosomiasis in two rural areas near Ismailia, Egypt. *Tropical Medicine and International Health*. 2, Supplement: A36–A47.

Mehanna, Sohair, Nadia H. Rizkalla, Hesham F. El Sayed and Peter J. Winch. 1994. Social and economic conditions in two newly reclaimed areas in Egypt: implications for schistosomiasis control strategies. *Journal of Tropical Medicine and Hygiene*. 97: 286–97.

Michelson, Edward H. 1993. Adam's rib awry? Women and schistosomiasis. *Social Science and Medicine*. 37: 493–501.

Miller, F. DeWolfe. 1978. Schistosomiasis in Rural Egypt. Manuscript Report. Cairo.

Miller, F. DeWolfe, M. Hussein, K. H. Mancy, M. S. Hilbert, A. S. Monto and R.M.R. Barakat. 1981. An epidemiological study of *Schistosoma haematobium* and *Schistosoma mansoni* infections in thirty-five Egyptian villages. *Tropical and Geographical Medicine*. 33: 355–65.

Moehlmann, Kristin. 2001. Girl-friendly schools improve Egypt's report card. www/unicef.org/information/mdg07.html, 5/09/01.

Moore, Henrietta L. 1988. *Feminism and Anthropology*. University of Minnesota Press: Minneapolis.

Morel, C. M. 2000. Reaching maturity: 25 years of the TDR. *Parasitology Today*. 16: 522–25.

Morley, David. 1993. The very young as agents of change. *World Health Forum*. 14: 23–24.

El Morshedy, H., M. Abou Nazel, A. Farghaly and H. Shatat. 1999. *Schistosoma mansoni* infection and cognitive functions of primary school children in Kafr El Sheikh. *Program and Abstracts of the 48th Annual Meeting of the American Society of Tropical Medicine and Hygiene. American Journal of Tropical Medicine and Hygiene*, 61 (3), Supplement: 243.

Morsy, Soheir A. 1978. Sex roles, power, and illness in an Egyptian village. *American Ethnologist*. 5: 137–50.

Morsy, Soheir A. 1993. *Gender, Sickness, and Healing in Rural Egypt*. Westview Press, Boulder, Colorado.

Nagi, Saad Z. 2001. *Poverty in Egypt: Human Needs and Institutional Capacities*. Lexington Books: New York and Oxford.

Nassar, Heba. 1996. *The Employment Status of Women in Egypt*. Report: Enhancing the Socio-Economic Status of Women in Egypt. Social Research Center, American University in Cairo: Cairo.

National Council for Women. 2000/1. Informational Pamphlet, in English. Cairo.

Nichter, Mark. 1994. Project community diagnosis: participatory research as a first step towards community involvement in primary health care. *Social Science and Medicine*. 19: 237–52.

Noda, Shinichi et al. 1988. Effect of mass chemotherapy and piped water on numbers of *Schistosoma haematobium* and prevalence in *Bulinus globosus* in Kwale, Kenya. *American Journal of Tropical Medicine and Hygiene*. 38: 487–95.

Nokes, Catherine, Stephen T. McGarvey, Linda Shiue, Guanling Wu, Haiwai Wu, Donald P. Bundy and Richard G. Olds. 1999. Evidence for an improvement in cognitive function following treatment of *Schistosoma japonicum* in Chinese primary schoolchildren. *American Journal of Tropical Medicine and Hygiene*. 60: 556–65.

Nooman, Z. M. et al. 2000. The epidemiology of schistosomiasis in Egypt: Ismailia Governorate. *American Journal of Tropical Medicine and Hygiene*. 62 (2), Supplement: 35–41.

Nunn, J. F. and E. Tapp. 2000. Tropical diseases in Ancient Egypt. *Transactions of the Royal Society of Tropical Medicine and Hygiene*. 94: 147–53.

Nutbeam, Don. 1998. Evaluating health promotion – progress, problems and solutions. *Health Promotion International*. 13: 27–44.

Ottawa Charter for Health Promotion. 2000. http://www.who.dk/policy/ottawa.htm

Parker, Melissa. 1992. Re-assessing disability: the impact of schistosomal infection on daily activities among women in Gezira Province, Sudan. *Social Science and Medicine*. 35: 877–90.

Parker, Melissa. 1993. Bilharzia and the boys: questioning common assumptions. *Social Science and Medicine*. 37: 481–92.

Partnership for Child Development. 1999. The cost of large-scale school health programmes which deliver anthelmintics to children in Ghana and Tanzania. *Acta Tropica*. 73: 183–204.

Poggensee, Gabriele, Hermann Feldmeier and Ingela Krantz. 1999. Schistosomiasis of the female genital tract: public health aspects. *Parasitology Today*. 15: 378–81.

Pope, R. T., B. L. Cline and M. A. Alamy. 1980. Evaluation of schistosomal morbidity in subjects with high intensity infections in Qalyub, Egypt. *American Journal of Tropical Medicine and Hygiene*. 29: 416–25.

Quarles, Wendy. 1993. Hygiene education – does it change behavior. In Cairncross and Kochar, *Studying Hygiene Behavior*, 145–53.

Rathgeber, Eva M. and Carol Vlassoff. 1993. Gender and tropical diseases: a new research focus. *Social Science and Medicine*. 37: 513–20.

Razum, Oliver, Regina Görgen and Hans Jochen Diesfeld. 1997. "Action research" in health programs. *World Health Forum*. 18: 54–55.

Richter, J. et al. 1996A. Reversibility of lower reproductive tract abnormalities in women with *Schistosoma haematobium* infection after treatment with praziquantel – an interim report. *Acta Tropica*. 62: 289–301.

Richter, J., Y. Wagatsuma, M. Aryeetey and H. Feldmeier. 1996B. Sonographic screening for urinary tract abnormalities in patients with *Schistosoma haematobium* infection: pitfalls in examining pregnant women. *Bulletin of the World Health Organization*. 74: 217–21.

Robert, C. F., S. Bouvier and A. Rougemont. 1989. Epidemiology, anthropology and health education. *World Health Forum*. 10: 355–364.

Rogers, Barbara. 1980. *The Domestication of Women: Discrimination in Developing Societies*. Kogan Page: London.

Ruf, Thierry. 1995. Histoire hydraulique et agricole et lutte contre la salinisation dans le delta du Nile. *Sécheresse*. 6: 307–17.

Saad, Reem. 1998. Hegemony in the Periphery: Community and Exclusion in an Upper Egyptian Village. In Hopkins and Westergaard, *Directions of Change in Rural Egypt*. 113–29.

Sadek, R. R. et al. No date. Comparative field study for *Schistosoma haematobium* diagnosis using reagent strips in Al-Minya Governorate, Egypt. Schistosomiasis Research Project: Cairo.

Sargent, Carolyn F. and Thomas M. Johnson (eds.). 1996. *Medical Anthropology: Contemporary Theory and Method*. Praeger: Westport, Connecticut.

El Sayed, Hesham, Sherif Abaza, Khadiga Dandash and Sohair Mehanna. 1998. The relationship between parasitic infections, anaemia and under nutrition among school children in rural Ismailia governorate, Egypt. *Zagazig University Medical Journal*. 4: 688–700.

El Sayed, Medhat K., Rashida Barakat, Azza Farghaly, Amal El Badway, Nina K. Soliman, M. Hassan Husein and F. DeWolfe Miller. 1997. The impact of passive chemotherapy on *Schistosoma mansoni* prevalence and intensity of infection in the Egyptian Nile delta. *American Journal of Tropical Medicine and Hygiene*. 57: 266–71.

Scott, J. A. 1937. The incidence and distribution of the human schistosomes in Egypt. *American Journal of Hygiene.* 25: 566–614.

Sholkami, Hania. 1990. *Sociocultural Factors Influencing the Prevalence of Diarrheal Disease in Rural Upper Egypt: an Ethnographic Study in Two Villages of Assiut.* UNICEF: Cairo and New York.

Sholkami, Hania M. 1988. "They are the government": Bureaucracy and Development in an Egyptian Village. M.A. Thesis, Sociology-Anthropology Department. The American University in Cairo.

Singerman, Diane. 1994. Where has all the power gone? Women and politics in popular quarters of Cairo. In Fatma Müge Göçek and Shira Balaghi (eds.), *Reconstructing Gender in the Middle East: Tradition, Identity and Power.* Columbia University Press: New York. 174–200.

Smith, Susan E., Timothy Pyrch and Arturo Ornelas Lizadi. 1993. Participatory action-research for health. *World Health Forum.* 14: 319–24.

Spencer, Harrison C, Ernesto Ruiz-Tiben, Noshy S. Mansour and Barnett L. Cline. 1990. Evaluation of UNICEF/Arab Republic of Egypt/WHO Schistosomiasis Control Project in Beheira Governorate. *American Journal of Tropical Medicine and Hygiene.* 42: 441–48.

Talaat, Maha. 2001. Impact of Schistosomiasis on Reproductive Health: Pilot Study. Final Report submitted to Social Research Center, American University in Cairo.

Talaat, Maha and F. DeWolfe Miller. 1998. A mass chemotherapy trial of praziquantel on *Schistosoma haematobium* endemicity in Upper Egypt. *American Journal of Tropical Medicine and Hygiene.* 59: 546–50.

Talaat, Maha and Mahmoud Omar. No date. Development and testing of methods of reaching children not screened and treated by schistosomiasis survey system. First Report for Special Program for Research and Training, School Task Force, WHO, Geneva, Switzerland.

Talaat, Maha, Mahmoud Omar and David Evans. 1999A. Developing strategies to control schistosomiasis morbidity in nonenrolled school-age children: experience from Egypt. *Tropical Medicine and International Health.* 4: 551–56.

Talaat, Maha, Afaf El Ayyat, Hanan Ali Sayed and F. Dewolfe Miller. 1999B. Emergence of *Schistosoma mansoni* infection in Upper Egypt: the Giza governorate. *American Journal of Tropical Medicine and Hygiene.* 60: 822–26.

El Tawila, Sahar. 1997. Child Well-being in Egypt: Results of Egypt's Multiple Indicator Cluster Survey. Report presented to UNICEF Egypt Country Office: Cairo.

UNDP. 1999. *Human Development Report 1999*. United Nations Development Programme and Oxford University Press: New York and Oxford.

Utzinger Jürg, Eliézer N'Goran, Amani D'Dri, Christian Lengeler and Marcel Tanner. 2000. Efficacy of praziquantel against *Schistosoma mansoni* with particular consideration for intensity of infection. *Tropical Medicine and International Health*. 5: 771–78.

van Ufford, Philip Quarles. 1993. Knowledge and ignorance in the practices of development policy. In Hobart, Mark (ed.) *An Anthropological Critique of Development: the Growth of Ignorance*. Routledge: London and New York. 135–60.

Vlassoff, Carol and Lenore Manderson. 1998. Incorporating gender in the anthropology of infectious diseases. *Tropical Medicine and International Health*. 3: 1011–19.

Wahba, Saneya. 1990. *Sociocultural Factors Influencing the Prevalence of Diarrheal Disease in Rural Upper Egypt: an Ethnographic Study in Two Villages of Aswan*. UNICEF: Cairo and New York.

WSSCC. 2000. *Vision 21: A Shared Vision for Hygiene, Sanitation and Water Supply and A Framework for Action*. Water Supply and Sanitation Collaborative Council, World Health Organization: Geneva.

Watts, Susan and Samiha El Katsha. 1997. Irrigation, farming and schistosomiasis: a case study in the Nile delta. *International Journal of Environmental Health Research*. 7: 101–13.

Watts, Susan and Samiha El Katsha. 1996. Women, schistosomiasis transmission and strategies for control: a case study in the Nile delta. *Journal of Environment, Disease and Health Care Planning*. 1: 17–27.

Watts, Susan and Samiha El Katsha. 1995. Changing environmental conditions in the Nile delta: health and policy implications with special reference to schistosomiasis. *International Journal of Environmental Health Research*. 5: 197–212.

White, Gilbert F. and Anne U. White. 1986. Potable water for all: the Egyptian experience with rural water supply. *Water International*. 11: 54–63.

White, N. J. et al. 1999. Averting a malaria disaster. *Lancet*. 353 (June 5 1999): 1965–67.

Whyte, Anne V. 1983. Community participation: neither panacea nor myth. In Bourne, *Water and Sanitation*. 221–41.

Wijeyaratne, Pandu, Eva M. Rathgeber and Evelyn St-Onge. 1992. *Women and Tropical Diseases*. International Development Research Centre, Ottawa, and UNDP/World Bank/WHO Special Progamme for Research and Training in Tropical Diseases, Geneva.

Wilkins, H. A. and M. El Sawy. 1977. *Schistosomiasis haematobium* egg counts in a Nile delta community. *Transactions of the Royal Society of Tropical Medicine and Hygiene*. 71: 486–89.

Wilson, R. A. and P. S. Coulson. 1998. Why don't we have a schistosomiasis vaccine? *Parasitology Today*. 14: 97–99.

World Bank. 2001. *Engendering Development: Through Gender Equality in Rights, Resources, and Voice*. A World Bank Policy Research Report. The World Bank and Oxford University Press.

World Health Organization. 2000. *World Health Report 2000: Health Systems, Improving Performance*. Geneva.

World Health Organization. 1998. WHO's Global School Health Initiative: Helping Schools to Become "Health Promoting Schools." Fact Sheet No. 92, revised June 1998. WHO, Geneva.

World Health Organization. 1993. *The Control of Schistosomiasis*. Second Report of the WHO Expert Committee, WHO Technical Report Series 830. Geneva.

Yousif, Fouad, Geoge Kamel, Mohamed El Emam and Shadia H. Mohamed. 1993. Population dynamics and schistosomal infection of *Biomphalaria alexandrina* in four irrigation canals in Egypt. *Journal of the Egyptian Society of Parasitology*. 23: 621–30.

Zou, Shimian. 2001. Applying DALYs to the burden of infectious diseases. *Bulletin of the World Health Organization*. 79: 267–68.

Zumstein, A. 1983. A study of some factors influencing the epidemiology of urinary schistosomiasis at Ifakara (Kilombero District, Morogoro Region, Tanzania). *Acta Tropica*. 40:187–204.

Index